FOREWORD

In striving to achieve sustainable economic development, each country has to make its own choices with regard to the energy sources needed to meet its objectives. In addition to assessing the available technology and the required resources, decisions should consider the true costs and benefits of pursuing alternative courses, taking account of their safety, their reliability, and their environmental and social implications.

As a contribution to the ongoing scientific and political debate on these questions, the international intergovernmental agencies have a major role to play in drawing together the accumulated experience and in providing an objective view of the related status of the different energy options. This report draws on some of the studies conducted by the OECD Nuclear Energy Agency in recent years, and in particular by its Nuclear Development Committee, and offers an overview of the current expert consensus on the technological and economic bases of nuclear power.

As such, the views expressed here do not necessarily correspond to those of the national authorities concerned. The report is published under the responsibility of the Secretary-General.

*

*   *

This report was prepared by Professor Peter M.S. Jones, Chief Economic Adviser to AEA Technology, United Kingdom, and former Chairman of the Nuclear Energy Agency's Nuclear Development Committee.

TABLE OF CONTENTS

Chapter 1

INTRODUCTION   7

What nuclear power can offer   7

Chapter 2

THE POTENTIAL DEMAND FOR NUCLEAR ENERGY   11

Historical background   11
Factors affecting future demand   15
Recent demand projections   17
Non-electric uses of nuclear power   18
Conclusions   19

Chapter 3

AN OUTLINE OF THE TECHNICAL BACKGROUND   21

Structure of the nuclear industry   21
Reactor technology   22
Uranium mining   24
Nuclear fuel manufacture   25
Reactor operation   25
Spent fuel management   29
Waste disposal   29
The overall nuclear fuel cycle   29

Chapter 4

THE AVAILABILITY OF NUCLEAR FUEL   33

Uranium   33
Other fuel cycle services   37
Services for advanced fuel cycles   40

Chapter 5

ECONOMIC COMPETITIVITY OF NUCLEAR POWER   41

The problems of cost comparison   41
Results of comparisons   42
The structure of nuclear generating costs   45
   a) Background   45
   b) Investment costs   45
   c) Nuclear fuel costs   47
   d) Non-fuel operating costs   48
   e) Decommissioning   49

The economics of plutonium fuels 51
Comparison with other electricity generation technologies 54

Chapter 6

GOOD PERFORMANCE AND FUTURE DEVELOPMENTS 57

Reactor development 57
   a)   General considerations 57
   b)   Improved reactor performance 57
   c)   Capital cost reduction 61
   d)   Collocation 63
   e)   Reduced construction time 64
   f)   Reduced commissioning period 65

Reactor Design 66
   a)   Advanced water reactors 66
   b)   Small and medium power reactors 67
   c)   High-temperature reactors 71
   d)   Advanced designs for improved fuel utilisation 71

Fuel cycle improvements 73
Overall performance 75

Chapter 7

TECHNOLOGICAL AND ECONOMIC CHALLENGES AND OPPORTUNITES 77

Present situation 77
Nuclear capital costs 78
Plant life extension 78
Decommissioning 79
The nuclear fuel cycle 79
Spent fuel stores and nuclear waste repositories 80
New technologies 80
Scientific exchange 81
Broader economic impacts 81
Constraints to nuclear development 83

Chapter 8

THE ROLE OF AN INTERNATIONAL AGENCY 85

Annex

NOTES FOR CHAPTER 5 87

Uranium fission 87
Discounting and rates of return on capital 87
Levelised cost calculations 87
Reprocessing economics 88

LIST OF ABBREVIATIONS AND GLOSSARY OF TERMS 89

BIBLIOGRAPHY 91

Nuclear Energy Agency publications 91
Nuclear Energy Agency studies in progress and forthcoming publications 91

Chapter 1

INTRODUCTION

What nuclear power can offer

The availability of abundant supplies of inexpensive energy has been the mainspring of economic advance over the ages. It underpinned both the industrial revolution of the 18th and 19th centuries and the rapid world economic growth of the 1950s and 1960s. On any objective criterion, the welfare of a large part of the world's population has been greatly enhanced in consequence.

The importance of energy and the critical dependence of industrial society on its continuing and continuous supply has been recognised for a long time. Vulnerability to potential constraints on or interruption of supplies has been a major consideration in national and international policy over the years, with concerns over long-term depletion of fossil fuel resources gaining prominence from time to time.

It is not surprising therefore that the discovery of nuclear energy with its vast potential was seized on by governments with alacrity, not least because it appeared to offer the prospect of attaining effective independence from fuel imports within a relatively short period.

In only 40 years, nuclear power has grown from a novel technology into a major world industry, with 17 per cent of the world's electricity supply now coming from nuclear reactors. Nuclear electricity supplies alone are now greater in quantity than the total world electricity supply available from all fuels in 1958.

However, there have been changes in the world outlook over the same period. In many countries nuclear energy development has been contentious. Some have welcomed and exploited it fully, some have been ambivalent, and some appear to have turned away from its future use altogether. It has been and remains a political issue at both the national and international level.

It is not alone in this. The growing demand for energy, even the need for industrial development and economic growth have also been the subject of critical questioning; albeit in those developed countries already enjoying the highest standards of living. It has long been recognised that industrialisation has not been achieved without cost, and that it has not benefited all nations and communities equally. The belated popular recognition in the affluent industrialised nations of the importance to mankind of avoiding environmental and ecological damage, has coincided with the growing awareness of the need to do something positive to improve the lot of the poorer nations, and with the latter's legitimate aspirations to break out of the poverty trap and enjoy more of the benefits made possible by modern industrial, medical and agricultural technology.

During this last decade of the twentieth century, the world is being confronted with problems of unprecedented significance. Until now man has had the ability to destroy himself, his neighbours and the ecology of smallish regions of the biosphere – and has often done so by his deliberate or inadvertent actions. In the future he could cause damage on a continental or global scale, largely as a consequence of growing populations and the sheer magnitude of human activity.

Conversely his power to alter things for the better is greater than ever. The tools and technologies exist to eradicate much of the world's poverty, famine and disease. The main barriers, apart from political will, are the economic constraints.

Some of the problems are readily discerned, but the solutions are harder to see; not least because we remain ignorant of many of the cause-effect linkages in the natural world and in human behaviour. The widely predicted environmental consequences of continued expansion in the use of fossil fuels remain speculations based on very imperfect mathematical models of the environment and the biosphere; even the existence, the causes and the effects of the phenomenon of acid rain were strongly contested only a few years ago. The greenhouse effect, to which carbon dioxide is the major man-made contributor, is an accepted scientific fact with a high degree of consensus on its future importance to world climate and environment, but the magnitude, timing, need for and urgency of countermeasures are hotly debated.

Nevertheless there is an imperative need, on current perceptions, to restrain fossil fuel combustion to levels with which the ecosphere can cope, and this does coincide with rising demand pressures from the growing populations of the developing nations, for the commercial energy without which their hopes are unrealisable. Even in the developed nations the potential additional uses of energy are far from exhausted, although improvements in energy efficiency may ameliorate their impact on overall demand growth.

Directly or indirectly, nuclear power, which unlike fossil fuel combustion produces neither corrosive and biologically damaging acid gases nor heat-trapping greenhouse gases, has the potential to make a significant contribution to the resolution of the impasse that would exist if carbonaceous fuels were the only available option. To do this, it has to satisfy governments and the public that it is a safe technology and that it is economic when compared with other large-scale supply alternatives.

Fortunately, because of its initial and rapid technological development for military applications, there was widespread international recognition and acceptance of the need to monitor and control nuclear power's use to ensure that the benefits were available to all, without the consequences that might follow from a proliferation of nuclear arsenals. The challenging nature of the technology and the need to contain safely the radioactive products and wastes were also recognised from the earliest days, although the standards set by and for the industry have become more demanding with the passage of time.

These features have led to continuing government involvement in all aspects of nuclear power development and the creation of international agencies and treaties, both for control and regulatory purposes, and to facilitate the exchange of technical information to the mutual benefit of participating countries. The International Atomic Energy Agency in Vienna and the OECD Nuclear Energy Agency in Paris have, in their separate roles, provided the foci for successful international technical and political collaboration.

That nuclear power can be a safe and economic energy source has been amply demonstrated. Nevertheless major problems have been caused by lack of human foresight and simple human error. The fact that nuclear energy has not been used as an offensive weapon since World War II is not an absolute guarantee that it will not be so used in the future. The public and politicians are right to be concerned about such matters and to take them into account in energy choices for the future.

If it were possible to satisfy humanity's aspirations without reliance on either fossil fuels or nuclear power, one at least of the global dilemmas would vanish. Few experts, if any, believe this possible, although optimistic scenarios based on high efficiency of energy use and maximum reliance on renewable energy sources, including biomass, have their advocates.

Each country has to make its own choices in the light of the existing, albeit imperfect, knowledge, and by taking account of the ongoing international scientific and political debate. This debate involves the future energy needs of developed and developing countries. It is concerned with the resources required and the technological means available to satisfy these needs. It is also concerned with the true costs and benefits of pursuing alternative courses, taking account of their safety, their reliability, and their environmental and social implications. However, choices by individual countries can no longer be assumed to have no impact outside their borders or be taken in isolation without consideration of their effects on others.

As a contribution to this debate, the international intergovernmental agencies have a major role to play in drawing together accumulated experience and in providing an objective consensus view on the technological and economic bases of the energy industries.

They have an obvious role in the provision of authoritative projections of energy demand and estimates of global mineral and renewable resources, so that the implications of alternative demand scenarios and fuel strategies can be analysed. This analysis can cover the feasibility of meeting demand, the actions necessary to avoid specific constraints (in manpower, finance and infrastructure as well as fuel supply), and the environmental consequences of alternative choices.

Costs to individual countries can be reduced and safety and reliability enhanced through co-operative research, through information exchange and technology transfer, conducted under the aegis of the international agencies. They provide a suitable focus for the definition and acceptance of international standards.

Nuclear energy is a technology where international collaborative technical and institutional development has been particularly important and fruitful. Indeed in its absence it is difficult to see how public and political confidence in the technology could be maintained and improved.

This report draws *inter alia* on the studies conducted within the ambit of the Nuclear Energy Agency in recent years, to present an overview of the current expert consensuses on nuclear technology, including the fuel cycle, and its economics; areas covered by the Agency's Committee for Technical and Economic Studies on Nuclear Energy Development and the Fuel Cycle (popularly abbreviated to the Nuclear Development Committee).

The specialised topics of non-proliferation controls, safety, radiological protection and waste disposal are not dealt with except insofar as they impact on the economics or technology of power generation and the fuel cycle. This is not to diminish their importance. Indeed public satisfaction on these aspects is crucial to the future development and wider application of nuclear power. Nevertheless, the nuclear industry itself is confident that these are aspects for which technical solutions exist. Solutions that will be backed, where necessary, by regulations and controls at the national or international level, to ensure that standards are seen to be rigorously maintained at all times.

Given that this confidence is justified and that it will be reflected in future public perceptions, there remains the question of what nuclear power can and may be expected to offer the world's energy and environmental policy-makers during the coming decades. This is the question addressed in the following chapters.

Chapter 2

THE POTENTIAL DEMAND FOR NUCLEAR ENERGY

Historical background

Since the nuclear fission process releases most of its energy in the nuclear fuel in the form of heat, it can be used as a source of direct heat or, through the use of steam turbines, as a source of electricity. It could also be used in combined heat and electrical power plants were this desired. The uses to which it has been and will be put are largely determined by economic considerations. These will be considered in Chapter 5. This chapter focuses on technical potential.

Some limited use has been and is being made of direct nuclear heat for industrial heating and other purposes, and nuclear reactors have been used for ship propulsion mainly, but not solely, for military vessels. In practice, however, the production of electricity has vastly outweighed other applications. Heating applications are widely dispersed with few large-scale geographically localised demand centres. Because of the high costs of transporting heat over long distances, heating plants have to be located very close to centres of demand, and will tend to be small to match demand. Both scale and location factors have in the past acted as barriers to the use of nuclear power for direct heating purposes. Electricity on the other hand is relatively cheap to transmit, and this encourages use of larger centralised generation plants that do not have to be built in close proximity to demand centres; factors which have favoured the use of nuclear power. These aspects will be considered in more detail later.

Studies of the demand for nuclear plant capacity have concentrated almost exclusively, therefore, on electricity production. Several such studies have been conducted under the auspices of the Nuclear Energy Agency. In common with most other economic and energy forecasting groups, the Agency's publications have erred historically on the side of optimism; in the Agency's case concerning the future levels of electricity demand and installed nuclear capacity (Figure 1). Rates of economic growth, and the consequential demands for energy in general and electricity in particular, have consistently failed to come up to earlier expectations. Additionally, the projected rate of penetration of nuclear power into electricity supply has not been realised, partly as a result of the reduced rates of new plant construction and partly due to the preference being given to other technologies in some countries.

The oil crises of 1973-74 and 1979-80, which were major disturbing factors, forced energy prices up and slowed world economic growth drastically. The concerns over energy availability and the rapid escalation of fossil fuel prices led many to expect a

Figure 1. **Projected nuclear capacity for OECD versus the year in which the projection was made**

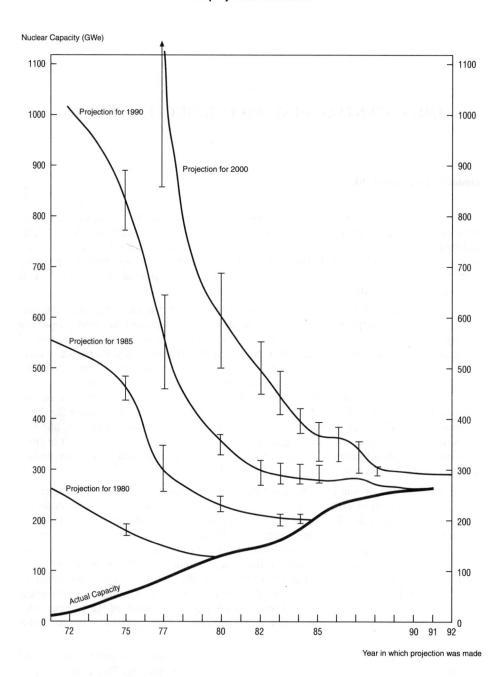

Source: *Nuclear Energy Data*, NEA, 1992

major swing to nuclear power. Indeed, plants were planned and uranium prices soared in the 1970s as utilities sought to contract for secure supplies.

However, the severe economic downturn in the early 1980s depressed electricity demand growth and deterred both general investment and investment in energy supply, so that a large part of the previously predicted capacity requirement did not materialise. Significant excess capacity developed in several countries as power plants (often oil or coal-fired) already under construction were completed. Future construction plans were delayed, and later, in many cases, cancelled. All types of plants were affected.

Only a few countries persisted and succeeded in their planned switch away from oil-dominated electricity supply. The United States, with by far the most ambitious nuclear programme, faltered. Uranium producers, whose exploration and production had expanded rapidly in anticipation of a continuing bonanza, saw prices tumbling as utilities themselves began to sell off their excessive stockpiles. The market recovery, widely projected for the early 1990s, has become a matter of uncertainty in the face of developments in the former Soviet Union and in eastern Europe: countries whose own nuclear programmes have been thrown into turmoil and who have themselves become vendors of uranium and some fuel services.

Several further factors have combined to weaken the nuclear industry's position further. Public opposition groups, acting through the courts in some countries, have delayed construction and operation of intended plants. This opposition has been strengthened by the accidents at the Three Mile Island plant in the USA and at Chernobyl in the USSR, as well as by continuing fears about the effects of radiation and the risks and consequences of accidents. Additionally, nuclear power's economic position has been squeezed by the effects of additional expenditures for safety and protracted construction delays (both arising in part from the previously mentioned concerns), and by the effects of plummeting fossil fuel prices from the mid-1980s. These, like uranium prices, have dropped as the quantities produced were boosted to meet anticipated demand growth which was, in the event, unrealised, so that vendors were faced with relatively depressed markets.

Financial barriers to investment in large-scale energy supply have also emerged in some countries as an unintended but direct consequence of government policies and institutional structures. In the United States, the regulation of prices by the State Public Utility Commissions has discouraged large-scale capital investment and created the phenomenon of rate-shock by delaying capital cost recovery on new investments until the plants entered production. The situation was aggravated by the delays caused by changing technical regulatory requirements and intervenor action in the courts. In the United Kingdom, the privatisation of the electricity supply industry, with the planned introduction of unfettered competition into power generation, has greatly increased the perceived financial risk of new investment, and will discourage private funding of large fossil-fuelled or nuclear plant construction, at least in the short term. In both countries the regulated low-financial-risk business of electricity production has become a much riskier investment, to the disadvantage of both nuclear and coal-fired electricity. In other countries too (e.g. Germany and Switzerland), the financial risks to utilities have been increased by intervenor actions.

Taken together, the effect of all these influences has produced a steadily declining expectation of nuclear power's contribution to electricity supply in the medium and long term. Indeed the fraction of electricity production obtained from nuclear power in OECD countries is expected to decline during the 1990s as the number of nuclear plant completions diminishes while electricity demand continues to grow, with electricity remaining the most buoyant sector of the energy supply scene. On a global scale, the picture has been made even gloomier by the decisions taken in the former USSR and other eastern European countries either to abandon construction of or to close down some nuclear plants in the aftermath of Chernobyl and the ensuing review of nuclear reactor safety throughout the zone.

This lengthy catalogue of negative influences is not wholly reflected in the most recent longer-term electricity and nuclear capacity demand forecasts released by the NEA, which are described later. On the other hand, the forecasts also take no account of possible developments arising from the current environmental concerns, particularly in relation to the greenhouse effect. These could have a major impact in the post-2000 period when much of the world's existing stock of generating plants, built since 1960, will be coming up for replacement.

Figure 1 illustrates how projections of installed nuclear capacity in OECD countries have declined since 1970. The picture is a depressing one for the industry which had, as indicated earlier, expanded rapidly to meet the anticipated requirements, both in terms of reactor design and construction capability and in terms of fuel production and fabrication capacity.

As a consequence of the economic pressures, the 1980s have been a period of rationalisation leading to the closure of many high-cost uranium mines, particularly in the United States and Canada, and to the withdrawal of a number of major companies or consortia from the reactor design and construction market.

The picture has not been one of unrelieved gloom however. Demand for nuclear fuel cycle services has grown steadily, in line with the growth of installed generating capacity. The nuclear share of electricity generation in the OECD as a whole had increased by the

Table 1. **Nuclear share of electricity supply in 1991**

Country	Nuclear electricity supply (percentage)	Share of electrical capacity (percentage)
Belgium	59.3	39.0
Canada	16.8	13.2
Finland	33.3	18.3
France	72.7	54.6
Germany	28.0	19.1
Japan	26.2	18.0
Netherlands	5.0	2.9
Spain	34.8	16.1
Sweden	51.6	28.7
Switzerland	38.8	19.4
United Kingdom	20.8	15.7
United States	19.4	13.7
Others	0	0
OECD total	23.5	16.2

Source: Nuclear Energy Data, NEA, 1992.

end of 1990 to 23 per cent, compared with 17 per cent for the world as a whole. Nuclear power has become an indispensable and major component of power supply in many countries both within and outside the OECD (Table 1).

Factors affecting future demand

Because of the relatively lengthy lead times associated with the construction of large conventional fossil-fired or nuclear-fuelled generating plants (five to eight years being typical), the levels of installed capacity up to the year 2000 are fairly narrowly defined. Because the slow-down in ordering new capacity in many countries during the 1980s affected nuclear plants more than other types, it seems inevitable, contrary to past expectations, that the overall nuclear share of electricity production will decline over the next ten to fifteen years. What will happen thereafter remains highly uncertain.

Barring global economic recession, there is wide consensus that the demand for electricity will continue to grow, both in the developed and the developing world. Factors determining the actual demand will include economic, demographic, technical and political developments, which interact in complex ways that preclude precise prediction, both of total demand and of the means of meeting it. The best that can be done is to rely on informed judgement to reach a view on a range of plausible scenarios for the future; scenarios that will normally exclude a whole spectrum of possible but, subjectively judged, improbable discontinuities.

Demographic factors are particularly important in the developing countries where the birth rate is high and the average age of the population low relative to those in developed countries. However, the pace at which they can develop their agriculture, infrastructure and industry depends not only on their own internal policies and efforts, but to a significant extent on the policies and economic performance of the developed nations as well, since the latter provide markets and are the principal sources of the technology and investment capital needed to underpin economic development.

Oil and increasingly natural gas are crucial to the world's transport and energy infrastructure. Both are subject to major regional supply-demand imbalances and both are potentially vulnerable to significant supply disruptions as a result of political events outside the control of the major consuming nations, as the events in the Middle East have repeatedly revealed.

Actual energy demand will be dependent on the efficiency with which it is used. The efficiency of energy utilisation has been increasing steadily, but future increases will be dependent on the existing infrastructure, on economic and policy factors, and on technical development, so that the required future levels of energy supply are uncertain even if end-use requirements could be predicted.

Environmental concerns are becoming an increasingly important factor in all policy areas, amongst which energy policy is prominent. On the energy supply side, the use of low-cost renewable sources like hydropower will be given priority, where available, although both it and the other less-developed renewable sources are not wholly benign and create their own environmental problems which, in some countries, have been judged to be unacceptable.

Pressures to reduce reliance on fossil fuel combustion may come to dominate energy policies, particularly in uses such as power generation where its reduction is potentially simpler to achieve than in transport applications. These pressures, which stem from concerns over the effects of acid rain and the potential effects of increasing atmospheric carbon dioxide levels on the future world climate, have grown considerably over the past few years. Further research might allay or exacerbate some of the current fears about the pace and impacts of climatic change, and either remove or reinforce the major environmental incentive to reduce dependence on coal, oil and gas. However it seems certain that prudent policies aimed at energy conservation and diversification away from carbonaceous fuels will be pursued in many, if not all, OECD countries.

Nevertheless, major changes in the world's energy infrastructure and attitudes to preservation of the biosphere will require a considerable degree of international political and popular consensus. This may prove hard to achieve in a manner that is seen by all to be equitable and practical within the prevailing pattern of economic constraints and national aspirations.

The developed industrial nations are better equipped technically and financially to make full use of advanced technological options like nuclear power. They have also, in most cases, exploited the low-cost options like hydropower to a considerable extent already. Many developing nations still have considerable potential for managed sustainable exploitation of their natural assets of biomass and hydropower, sun and wind. On the other hand, this is not always the most cost-effective option, and the use of fossil fuel offers the most rapid route to improved lifestyles in countries like India, China and Indonesia. In such countries, and many others in Asia, Africa and the Americas, nuclear power is expected to grow in importance, but practical constraints severely limit the rate at which it could substitute for fossil fuels even if this were a global environmental objective. These constraints include the need for an adequate technical infrastructure, access to trained manpower and the availability of finance.

This leads to the question therefore of whether the developed countries within and outside the OECD will consider it desirable to accelerate their switch away from fossil fuels beyond the rate of their natural economic replacement as a contribution to global goals that impose other, but no less significant costs on the poorer countries of the world.

It is impossible to do more than raise these possibilities at this point in time, but it will be evident that current awareness of problems, be they real or imagined, precludes confident forecasts of energy demand and supply very far beyond the immediate planning horizons of perhaps five to ten years. The need for caution over even this limited timescale is reinforced by the experience of attempted projections in recent decades (Figure 1).

Nevertheless, projections are wanted by the energy industries and by governments. The location and development to the exploitation stage of new mineral resources is time-consuming, with periods of ten or fifteen years not uncommon. Large-scale conventional coal-fired or nuclear power plants can take up to ten years to plan and bring into operation. They have economic lives of, perhaps, forty years; longer if refurbishment is undertaken. Tidal barrages and major hydropower schemes are not very different. The use of smaller simpler installations such as gas turbines (combined-cycle or peaking plants), small fossil-fuelled co-generation plants and wind turbines may help to reduce the short-term supply planning problem, but may sacrifice economic and scale benefits and could prove unexpectedly expensive if unanticipated fuel supply or maintenance problems arise.

There are clearly advantages in having energy futures reviewed by international agencies where a spectrum of different views, expectations and concerns can be brought together to forge a better informed, though not necessarily correct, consensus.

The International Energy Agency and the Nuclear Energy Agency, both within the OECD, provide such fora, as does the International Atomic Energy Agency. The Nuclear Energy Agency has taken the lead in producing demand projections for nuclear power based on member nations' national expectations to 2010 and on simple projection models thereafter.

Recent demand projections

The most recent projections are illustrated in Table 2. The range of the longer-term nuclear electricity projections specifically excludes major deviations from present trends, such as those that might follow a nuclear moratorium or a general policy of an accelerated switch to nuclear power, for environmental or other reasons. Economic growth rates of 2.6 per cent per annum were assumed for OECD countries and 3.2 per cent per annum for the non-OECD WOCA countries (WOCA, or the World Outside the Centrally Planned Economies Area, is no longer an accurate term following the changes in eastern Europe, but it is retained here, following the original data set, to define the countries that were not in the East-European Bloc and China. Statistics for the latter countries have not been readily available). These rates were consistent with long-term growth patterns, but they presuppose no return to the accelerated growth the world economy experienced during the 1950s and 1960s.

Nuclear electricity shares of total supply were assumed to rise towards, but not reach, asymptotes of 20 or 25 per cent in North America and 40 per cent or 60 per cent in OECD Europe and Japan, thus generating the high and low nuclear electricity and capacity projections shown in Table 2.

Table 2.   **Projected electricity demand and nuclear supply in OECD countries**

Year	Electricity demand TWh	Nuclear supply TWh	Nuclear share (percentage)	Nuclear capacity GWe
1990	6 650	1 530	23.0	262
2000 [a]	7 840	1 827	23.3	300
[b]	8 457	1 770	20.9	292
2010 [a]	9 250	2 320-2 600	25.1-28.1	365-414
[b]	10 330	2 020	19.6	234
2020 [a]	10 680	2 800-3 500 ·	26.2-32.8	430-563
2030 [a]	12 190	3 345-4 390	27.4-36.0	496-651

a) NEA projections from *Electricity Generation from Nuclear Reactors and Uranium Demand to 2030*, NEA, 1990.
b) Sum of OECD country projections from *Nuclear Energy Data*, NEA, 1991.

Since these longer-term projections were produced, there have been policy decisions in several countries, including Canada, Spain, Switzerland and the United Kingdom, that have resulted in de facto moratoria on further nuclear plant construction, at least in the near term. Finland, the Netherlands and Sweden on the other hand appear, at the time of writing, more favourably disposed to nuclear power than they were. Some of these changes have already influenced the short-term projections (to 2010) assembled from national returns and might contribute to a reappraisal of the longer-term future position when the projections are next revised. The divergence between the two sets of medium-term projections is apparent in Table 2.

Non-electric uses of nuclear power

Separate consideration has been given from time to time to the possible expansion of non-electrical uses of nuclear power. Such uses have been demonstrated on a commercial scale for steam heating for production of heavy water, for district heating of domestic and commercial premises, for desalination of sea-water to produce fresh water, and for marine propulsion of submarines and surface vessels including ice-breakers.

These specific applications and many more can be satisfied using the steam temperatures achieved in pressurised water-cooled reactors. A large proportion of industrial heating uses, as well as general building heating, fall into this category.

Attention has also been given, however, to higher-temperature applications for which light-water reactors are not suited. These include thermochemical hydrogen production, synthetic natural gas production and steel making. The capacity demands for these more specialised applications are limited, and they have so far not justified the development and construction of a full-scale demonstration reactor and plant. Successful experimental helium-cooled high-temperature reactors capable of supplying heat at the required temperatures have however been operated and have their advocates. They have specific technical attractions for wider applications (see Chapter 6).

In principle, if a nuclear reactor can provide electricity as cheaply as fossil-fuelled sources, it must also be able to provide a cheap source of heat. However heat, in the form of high-temperature steam or hot water, is expensive to transport for any distance through the necessary insulated pipes, so that regardless of the fuel used, heat production plants, unlike electricity generating plants, need to be located in close proximity to the sites where heat is needed.

This has two consequences for nuclear power. First, most localised heat demand is insufficient to require plants of the size now conventionally deployed in OECD countries, and secondly, the regulations concerning the siting of light-water reactors, which are the reactor types in most widespread use, require their location away from major centres of population. Additionally, capital-intensive nuclear plants benefit economically from being used intensively with a normal output approaching their full design capability. Stable demand is characteristic of some industrial processes but not of building heating where there are major seasonal variations.

The need for assurance of continuity in most heat supply applications also means that local backup capacity is needed to cover any periods of planned or unplanned shutdown of the reactor. This can be provided by employing a number of smaller nuclear plants, each supplying a small portion of the load, or by alternative backup technologies.

Either solution is expensive. For these reasons, nuclear power is not ideally matched to general heat-only applications, although the same problems are not necessarily associated with applications where a reactor is tied to specific propulsion or industrial applications that can be interrupted, when the need arises, without difficulty. Thus, provided nuclear plant reliability is high and unplanned outages are few, the need to shut down a reactor for refuelling need not be a deterrent to its use.

In situations where only a small part of a reactor's heat is used directly, the remainder being used for electricity production, the economics may be more favourable. The electrical output can be varied to employ the surplus heat, thus providing complete flexibility in heat supply (always supposing that there is a use for the electricity). In other situations, low-temperature "waste heat" from an electricity generating plant may be used to heat buildings, for example.

Unless nuclear power can penetrate markets for heat and propulsion, its overall contribution to world energy supplies will remain limited. Until recently none of the non-electrical outlets has appeared likely to provide a large market for nuclear power in the foreseeable future. The economics were regarded as unattractive, because nuclear plants were seen as benefiting from large-scale construction. However, recent enhanced interest in smaller plants (see Chapter 6) could provide a route, if and when they are fully developed and demonstrated, to widen the markets for nuclear power into countries with smaller electrical grids and into direct-heat applications. The environmental incentive to replace fossil fuels may also give a stimulus to such changes and, ultimately, there is no reason why nuclear energy should not satisfy the bulk of the world's energy requirements directly or indirectly.

The indirect routes would be via the substitution of hydrogen (or methane or methanol) produced using nuclear energy, either electrolytically or thermochemically, for conventional fossil fuels in heating and transport applications.

The total markets for nuclear power in this area have been examined, but they remain market opportunities rather than expectations, since they could only materialise given conscious acts of development and application which call for a clear policy objective. Speculative development in this area seems unlikely without significant government backing in one or more major countries; backing which is not yet in prospect.

Conclusions

From what has been said so far, it will be apparent that the future of nuclear power remains very much a matter of choice. This choice will be determined largely by the decisions taken by governments acting alone or in concert. Their decisions will inevitably be influenced by public attitudes and concerns on both nuclear power and the other energy supply options.

The potential primary users of nuclear energy, the electricity (and in the future heat) supply industries, will make their choices on the basis of least-cost routes to meeting projected demand within the planning, regulatory and institutional framework created by governments.

This framework may be designed with energy as a central focus. Industrial actions may be controlled or influenced either entirely by regulation backed by inspection, or they may be determined by market forces as influenced by government fiscal and environmental policies. In the latter case, the deciding factors may not be directly or wholly related to energy supply and use.

For nuclear power to fulfil its full potential, it will have to satisfy the public and politicians of its safety and its ability to deal with its wastes. It will also have to maintain a competitive position economically relative to other energy sources (see Chapter 5).

The post-2000 projections presented in Table 2 represent a range of views of how nuclear electricity and nuclear capacity might grow if the world economy and government policies carry on much as they have done over the past decade. They could be radically altered up or down by events or policy decisions that tended to favour or disfavour economic growth, electricity use or nuclear power itself.

In the longer term, nuclear power's prospects will depend to a major extent on technological developments; notably in relation to reactors that are able to utilise uranium resources efficiently and reactors that facilitate nuclear power's penetration of non-electric energy markets. More is said on these aspects in Chapter 6.

Chapter 3

AN OUTLINE OF THE TECHNICAL BACKGROUND

Structure of the nuclear industry

The nuclear industry is frequently thought of as having two main sectors, a power plant industry and a fuel cycle industry. In reality, however, there are six different sectors within the industry, half of which arise from the specific technical characteristics that serve to distinguish it from the other energy supply industries.

The sectors are uranium mining, nuclear fuel manufacturing, reactor design and construction, power supply and reactor operation, spent fuel management, and finally radioactive waste disposal. In practice some of the sectors have grown up in multi-functional companies so that fuel manufacture, spent fuel management and radioactive waste disposal are operated as separate activities within a single organisation in some countries.

The nuclear industry differs from the fossil-fuelled electricity industry in that its fuel is a precision-engineered product, and it goes to great lengths to contain its waste products and keep them away from the biosphere and environment. This contrasts with the fossil-fuelled electricity industry which is based on use of unprocessed naturally occurring materials and which has, until recently at least, used the environment as a free waste repository. (It is true that oil is refined to produce a wide range of fuels and other products, but the heavy fuel oil used in the majority of large-scale oil-fired power stations is a residual rather than a primary product of refining). In recent years, fossil fuel combustion has become the focus of great attention and, in many countries, techniques have been or are being introduced as an integral part of the combustion process to reduce the emissions of acid gases. These give rise to liquid waste, sludges, or solid residues which themselves have to be disposed of. No techniques are yet available to contain carbon dioxide, the principal man-made greenhouse gas, which is an inescapable product of fossil fuel combustion, although possible routes are being explored, including removal of equivalent amounts of carbon dioxide from the atmosphere by means of tree-planting programmes.

In the following paragraphs, a brief sketch of the technical background is given for each of the component sectors of the nuclear industry to underpin the discussion in later chapters. The descriptions commence with reactor technology, since the choice of reactor determines the nature of the fuel.

Reactor technology

A nuclear reactor is a device in which atoms of a nuclear fuel (normally uranium or plutonium) are split (fissioned) by neutrons to yield large quantities of heat, together with a range of lighter highly radioactive atoms (fission products) and around three further neutrons. These neutrons can go on to fission additional atoms of fuel, or they may be captured without producing fission by fuel atoms or the structural materials of the reactor.

The neutrons released in the fission process have high speeds (fast neutrons) and are relatively inefficient at fissioning the normal reactor fuel, uranium. They are therefore slowed down, using materials called moderators, to speeds similar to those of molecules in the air around us to become so-called thermal neutrons. Reactors relying on uranium fuels and employing moderators to slow down neutrons are known as thermal reactors. Some reactors do not contain moderators and rely on fast neutrons to produce fissions in the fuel. These are the so-called fast reactors. Such reactors normally rely on plutonium fuels; plutonium being a material produced when neutrons are captured by uranium atoms in the fuel without inducing their fission.

Reactors of either type, i.e. thermal or fast, employ a coolant to remove heat from the fuel. The coolant provides the means by which the heat generated in the nuclear fission process can be utilised for power production or for other heating purposes.

The rate at which the fissions take place in the reactor can be controlled by the introduction of neutron-absorbing materials in the form of rods (control rods), which can be raised or lowered within the core of the reactor where the fuel is positioned. Fine-tuning can also be effected by controlling the temperature of the coolant within the reactor core. The control rods can be used to stop the fission process completely and thus shut the reactor down for maintenance, for refuelling, or in the event of an emergency.

Even when a reactor is shut down, the irradiated fuel in its core continues to produce large amounts of heat from the decay of the highly radioactive fission products. It is therefore essential to reactor safety that coolant flows are maintained at all times to prevent the fuel from melting and damaging the reactor.

The fuel itself is contained in sealed metal tubes that are designed to contain the highly radioactive fission products, both while the fuel is in the reactor and subsequently. The reactor core, with its moderator and coolant, is in turn contained within a sealed vessel that is surrounded by a thick biological shield designed to ensure that the workers are protected from the intense radiation produced within the core.

A number of parallel lines of reactor development, each with its own logic and advocates, were pursued by the newly created nuclear power industry. There was no clear a priori reason to believe that any one reactor type had a clear economic or technical advantage over its competitors, and national considerations of access to fuel, moderator and fuel services weighed heavily in initial national and commercial preferences.

The main lines of development relied on water, heavy water or graphite as moderators to reduce fission neutron energies, and water, heavy water or gas as coolant to remove fission heat. Steam produced by the reactor, either directly by boiling the water coolant in the reactor core, or indirectly using the coolant to boil water in heat exchangers, drove the turbines used to generate electricity.

The majority of countries have now adopted the water-cooled and water-moderated systems as their mainstream nuclear power plant in the form of the pressurised light-water reactor (PWR or LWR), the boiling-water reactor (BWR) or the pressurised heavy-water reactor (PHWR). The first of these now dominates the market, providing 63 per cent of existing nuclear capacity. BWRs provide 26 per cent and PHWRs 6 per cent.

Graphite-moderated gas-cooled reactors were initially favoured in France and the United Kingdom, but both countries have subsequently switched to pressurised light-water reactors on the basis of their perceived economic advantages (the gas-cooled Magnox and Advanced Gas Reactors are still the basic nuclear power sources for the United Kingdom electricity system however). Gas-cooled reactors remain a topic of research interest, particularly for high-temperature applications, but there are no commercial-scale plants planned at present.

The Soviet Union deployed a hybrid water-cooled graphite-moderated reactor (acronym RBMK) on a significant scale, but the Chernobyl accident has highlighted the inherent problems of this design as it was conceived and operated in the USSR, and future thermal nuclear plants are likely to be built using their pressurised water reactor design (with the acronym VVER).

The only other reactor type that has received significant attention has been the liquid-metal-cooled, unmoderated fast neutron reactor, which has the potential to breed its own fuel from otherwise useless depleted uranium. This reactor has been brought to the

Figure 2. **Pressurised water reactor (PWR)**

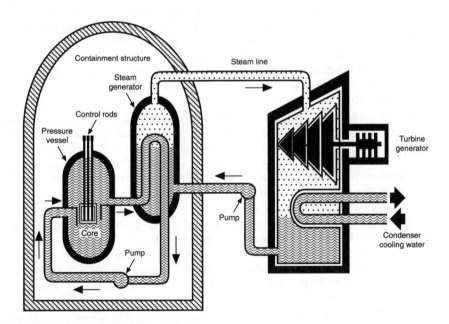

Source: Decommissioning of Nuclear Facilities, NEA, 1991

prototype stage in several countries and France has built one commercial-scale unit (Superphenix).

Each reactor type has evolved and new reactors have benefited from the practical experience in the design, construction and operation of earlier plants employing the same, or indeed different, basic technologies. Designs have also been adapted to take account of changing regulatory requirements.

For the most part, steady progress has been achieved in reducing unplanned shutdowns, improving plant availability and output, reducing operator radiation exposures, and overcoming the problems caused by materials degradation in the hostile environment of the reactor core (see Chapter 6).

Although the Nuclear Energy Agency has regularly reviewed the characteristics of the different reactor types in connection with the influence their deployment could have on uranium demand and utilisation, most economic and technical studies have concentrated on the dominant water-cooled reactors, the PWR and BWR, since these have the widest interest.

The studies have examined routes to good performance, including reduction of capital cost, the status of and specifications for the development of advanced thermal reactors, and the decommissioning plans and cost expectations of OECD countries.

Uranium mining

Uranium is the material which provides the fuel for the bulk of the world's nuclear reactors. Natural geological processes have produced workable concentrations in a wide variety of mineral formations. These range from vein-type deposits where the uranium-containing minerals have been forced into cracks in a host rock, to sandstone deposits where the uranium has been concentrated in sediments by favourable chemical processes. Although some deposits contain remarkably high concentrations of uranium amounting to well over 10 per cent by weight, the majority of economically workable ores contain only around 0.2 per cent.

A wide range of prospecting techniques is used to locate promising ore bodies, whose presence is then confirmed by chemical analysis and its extent delineated by drilling boreholes. Where deposits look promising, drilling will continue in order to establish the level of recoverable uranium reserves in the ore body and whether they can be recovered economically. If initial results are discouraging or the economic outlook is unfavourable, the ore body may not be fully investigated.

The uranium-bearing minerals may be worked by open-cast mining methods or by conventional deep mining, or, in some cases, the uranium is recovered by leaching methods in which chemical solvents are pumped through ore-bearing strata to recover the uranium in solution. Mined ores are crushed in a processing plant (or mill) and the uranium extracted by chemical methods appropriate to the composition of the mineral concerned. Some 85 to 95 per cent of the uranium will normally be recovered as a solid concentrate called yellowcake.

Typically, for each tonne of uranium extracted, some 500 tonnes of rock and ore would have been crushed and treated. The residues, called mill tailings, together with the treated waste liquids from chemical processing, have to be disposed of in ways that minimise the risk of transfer of uranium and its decay products to groundwater. Solid tailings piles, for example, can be covered over with soil to help retain the radioactive radon gas produced by natural radioactive decay of minerals in the tailings and to screen off gamma radiation. A great deal of attention has been paid in recent years to the problem of safe management of these tailings in order to ensure that environmental and potential health impacts are minimised. Less stringent precautions were taken in the early days of the uranium industry, before the risks were fully appreciated and the current regulatory frameworks established in OECD countries.

Nuclear fuel manufacture

The uranium concentrate, yellowcake, is purified and converted to uranium dioxide, a stable chemical compound with a high melting point, which is used as the fuel in the majority of the world's reactors. Pressurised heavy-water reactors use uranium dioxide fuel made directly from natural uranium, but the predominant light-water reactors and the United Kingdom's advanced gas-cooled reactors require uranium dioxide that has had the concentration of the fissile uranium isotope, uranium-235, increased from its natural concentration of 0.7 per cent to around 3 per cent.

This concentration or enrichment of the lighter isotope relative to the naturally predominant isotope (uranium-238) involves conversion of the uranium to a volatile uranium compound, uranium hexafluoride, and its physical processing in either a diffusion plant (United States and France) or a high-speed centrifuge plant (Germany, the Netherlands and the United Kingdom). The uranium hexafluoride enriched in uranium-235 is reconverted to uranium dioxide for use in the fuel, while the fraction depleted in uranium-235, known as the enrichment plant tails or depleted uranium, is stored for possible future use as a source of fuel for fast reactors. (There are other enrichment processes in use outside the OECD countries, and new processes are being developed within the OECD involving the application of lasers, but these are still some way from being commercialised).

The uranium dioxide fuel is compressed into ceramic pellets, which are then stacked in sealed metal tubes (cans or cladding) to produce fuel pins. A number of these pins are held rigidly together in a lattice of fixed geometry to make a fuel assembly, which is transported to the reactor for use as fuel.

The composition of the fuel, including its enrichment, the material used for cladding, the dimensions of the fuel pins and the number of pins in an assembly are dependent on the reactor in which the fuel is to be used. The enrichment may also vary depending on the intended location of the fuel in the reactor core and the stage in the reactor's life at which the fuel is to be introduced.

Reactor operation

Reactor operation is the responsibility of the electrical utility. It includes a number of stages, from the loading of the initial uranium fuel into the reactor core and reactor commissioning, through the routine operation stages with the provision of power to the

One of these reactor fuel assemblies can produce as much electricity as approximately 3 500 tonnes of coal in a conventional power plant.

Credit : Kraftwerk Union, Germany

Credit : Cameco Corporation, Canada

The Deilmann open-pit uranium mine in Saskatchewan, Canada.

Credit : COGEMA, France

Reactor spent fuel can be safely stored for many decades. Above, the spent fuel storage pool at the La Hague reprocessing plant, in France.

Credit : EdF, France

Credit : PNC, Japan

Cost reductions of some 20 per cent or more have been achieved by building a number of reactors on the same site.

By using "fast reactors", the quantity of energy recovered from natural uranium can be increased by some 60-fold. Shown above is the MONJU prototype fast reactor, in Japan, which is scheduled to begin operation in 1993.

electricity grid, up to the final shutting-down of the reactor when it is taken out of service and its ultimate decommissioning.

During the period of routine operation, the utility will control the heat output in the reactor core, within the range for which the reactor was designed, taking account of the power needs of the grid. Because their fuel and operating costs are low, nuclear plants in most countries are used to the maximum extent possible so that they supply electricity at a steady rate to meet the grid's baseload requirements. In some countries, where the nuclear share of total generation capacity is very high, a need can arise to operate nuclear plants at reduced loads and to vary the output to follow the load requirements imposed by the grid.

Nuclear fuel has a typical residence time within the reactor core of around three years, although shorter periods are demanded of some of the initial fuel in the first few years after a reactor has been commissioned, and of some of the final fuel in the few years immediately prior to a reactor's final shut-down. A part of the reactor's old (or spent) fuel is removed at roughly annual intervals, timed to coincide with periods of low electricity demand wherever possible, and replaced with fresh nuclear fuel. For the light-water reactor, this refuelling requires that the plant be shut down and taken out of service. Pressurised heavy-water reactors and some gas-cooled reactors can however be refuelled while operating and do not have to be closed down.

During the periods of refuelling, the operators may take the opportunity to redistribute fuel in the core to maintain the most efficient operation, and they will undertake any other necessary maintenance or refurbishment of reactor components. Statutory safety inspections of the reactor itself will also be phased in with operations calling for routine shut-down. The reactor operators make every effort to ensure that the reactor is in operation as continuously as possible, consistent with the need for refuelling, and every effort is made through planned maintenance and intensive quality control to ensure that unscheduled shut-downs arising from technical problems are kept to a minimum.

Another major consideration is the need to minimise the radiation doses received by reactor maintenance staff during periods of refuelling, maintenance or refurbishment. To this end, operators are continuously reviewing and working to improve the procedures and to cut down the time required for these operations.

Few commercial reactors have yet reached the end of their economic lives. When they do, they are shut down and the process of decommissioning the plant can commence. This process is normally considered to consist of three phases. The first is the removal of spent fuel, the second the dismantling of the non-nuclear plant and buildings associated with the reactor, and the third the dismantling of the highly radioactive reactor core itself. The three stages may be separated in time. The defuelling operation would commence immediately after reactor closure, whereas the stage-two dismantling can be conducted at a more leisurely pace over an extended interval. The timing of the third decommissioning stage involves careful consideration of the merits of early removal of the radioactive plant, balanced against the unquestionable advantages of deferring action until the induced radioactivity in the core structure and pressure vessel has had a chance to decay naturally to much lower levels. This decay reduces the radiation risks for the workforce engaged in the ultimate dismantling operations.

Spent fuel management

After a period, the fuel depleted in the fissile uranium-235 isotope (spent fuel) is removed from the reactor and stored under water in cooling ponds at the reactor site. After its initially very high radioactivity has decayed sufficiently, it may be removed from the ponds for monitored storage away from the reactor or transferred to a reprocessing plant. In this plant, the spent fuel is chemically dissolved to recover plutonium and unburnt uranium, including the residual uranium-235, from the highly radioactive mixture of fission products and unwanted heavy elements. These fission products and heavy-metal wastes (actinides) are the so-called highly active wastes, which are stored in solution until they can be incorporated into a suitable solid matrix which, in due time, will be disposed of in underground repositories. At the present time, this type of waste is being incorporated in a glass matrix (vitrified) at reprocessing plants in France and the United Kingdom.

The recovered uranium and plutonium can be recycled as fuels for thermal or fast reactors after suitable processing.

The bulk of spent fuel is currently stored in cooling ponds at the reactor site. A number of other techniques for prolonged storage, including dry storage methods, have however been under development. Some of these could be used for storage away from the reactor site and this could become a significant new business in due course.

Not all countries plan to reprocess spent fuel. The so-called ''once-through'' nuclear fuel route envisages that spent fuel will be conditioned, placed in sealed containers and disposed of in the same way as high-level radioactive wastes.

Waste disposal

The disposal of nuclear wastes, particularly the high-level radioactive wastes arising from spent fuel reprocessing (or spent fuel itself if not reprocessed) has proved a difficult issue for governments in many OECD countries.

All countries necessarily have sites allocated for the shallow burial of the low-activity wastes arising from the nuclear industry, and from medical and industrial operations. But few, with the notable exception of Sweden, have yet formulated clear policies and identified and developed sites for the deep geological disposal of wastes.

A great deal of research has been in progress on the suitability of different geological media for the deep burial of wastes or spent fuel, and the scientific community is satisfied that proven techniques exist to ensure the safe containment of long-lived wastes. There is also confidence in the ability of the nuclear industry to demonstrate that sites selected as waste or spent fuel repositories will ensure that the risks to future generations are negligible.

The overall nuclear fuel cycle

The nuclear fuel cycle illustrated in Figure 3 is the commonest currently in use. There are two distinct branches, depending on whether it is planned to reprocess spent fuel or dispose of it without reprocessing. There are variants of this scheme matched to

29

Figure 3. LWR fuel cycle - material flows

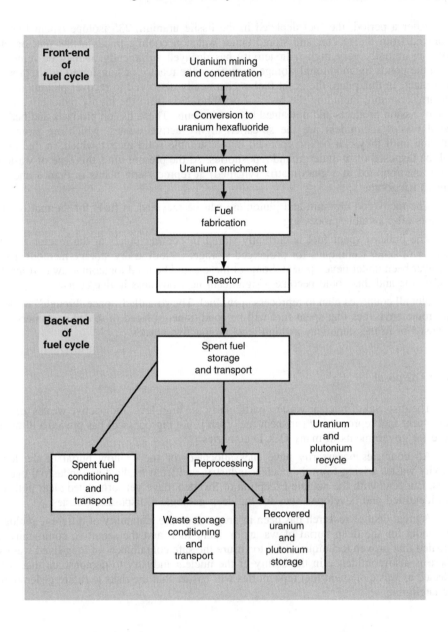

Note: 3000 tonnes of ore gives 6 tonnes of uranium, which yields 1 tonne of enriched uranium dioxide fuel. 1 tonne of nuclear fuel produces 0.27 terawatt-hours of electricity. 1 tonne of spent fuel yields 0.96 tonnes of uranium, 9 kg of plutonium and 30 kg of fission product wastes.

Source: The Economics of the Nuclear Fuel Cycle, NEA, 1985

different reactor types. Thus, pressurised heavy-water reactors such as Canada's CANDU use unenriched natural uranium oxide fuel, the UK's Magnox gas-cooled graphite-moderated reactors employ metallic natural uranium fuel, and fast reactors and some light-water thermal reactors use mixed uranium and plutonium oxide fuels known as mixed oxides, or MOX. Other more specialised fuels have been developed for particular reactors such as the high-temperature gas-cooled reactor. Some of these fuels are discussed later in the text, but they do not have immediate relevance to the more general question of fuel availability, which is the subject of the next chapter. This is partly because they are not, for the most part, significant in quantitative terms. They are usually produced in special plants associated with and matched to the particular reactor development programme, and they are not internationally traded on a significant scale.

More details of the nuclear fuel cycle and the relevant technologies can be found in a number of standard works and in the NEA publications listed in the bibliography.

Chapter 4

THE AVAILABILITY OF NUCLEAR FUEL

Uranium

Uranium is widely distributed in the earth's crust and present in low concentrations in sea water. Nevertheless, economically workable deposits are not plentiful and those that exist are largely concentrated in a small number of countries. Because it occurs in many mineral forms and the size of deposits is very variable, the costs of its extraction and purification differ considerably from source to source.

A standard classification of resources, in terms of their recovery costs and the degree of confidence in the estimates of their quantitative extent, has been developed as a basis for the regular definitive reviews of resources conducted by the OECD Nuclear Energy Agency jointly with the International Atomic Energy Agency.

The so-called "known" resources estimated to be recoverable at less than $130/kg of uranium (US$ 1990) amount to some 3.6 million tonnes, of which 2.5 million tonnes are estimated to be recoverable at under $80/kg of uranium (U). These resources are made up of well-delineated reasonably assured resources (RAR) and less well-delineated estimated additional resources (EAR-I), both of which are predominately located in Australia, North America and South Africa (Table 3 and Figure 4).

Table 3. **Uranium resources in OECD and developing countries (WOCA)**

Million tonnes U

Category	Cost range		
	up to $80/kg U	$80-130/kg U	$130-260/kg U
Reasonably assured	1.55	0.65	0.40+
Estimated additional I	0.77	0.39	0.19+
Estimated additional II	0.67	0.52	0.65+
Speculative	9.6-12.1*		0.5**

+ Reported levels.
* Based on *International Uranium Resources Evaluation Project* most likely range.
** Based on levels reported to OECD.
Source: Uranium Resources, Production and Demand, NEA, 1990.

Figure 4. **Distribution of uranium resources among WOCA countries**

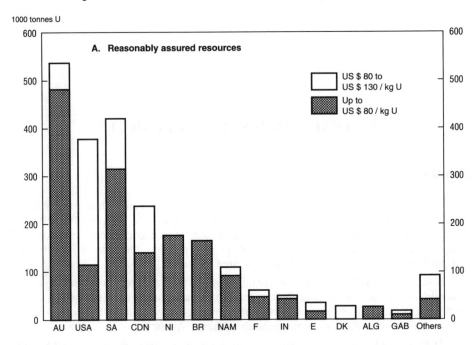

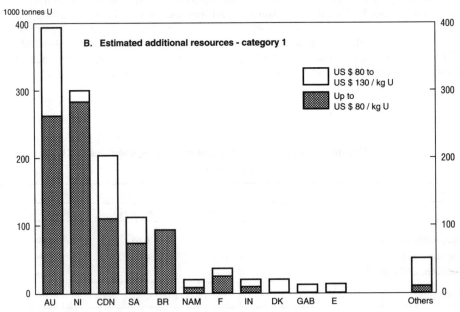

Note: AU = Australia, USA = United States, NI = Niger, SA = South Africa, CDN = Canada, BR = Brazil, NAM = Namibia, F = France, IN = India, DK = Denmark, E = Spain, GAB = Gabon, ALG = Algeria

Source: Uranium Resources, Production and Demand, NEA, 1990

Limited estimates exist for the quantities of known resources recoverable at costs between $130 and $260/kg U, but these resources are too expensive to exploit in current market conditions so that there has been little interest in extending knowledge in this area.

All these figures have conventionally excluded the resources of the non-WOCA countries (the former USSR, eastern Europe and China) for which detailed data have not been available, in the same way as they have been excluded from demand forecasts.

If the "known" resources were all that existed, the contribution of thermal fission reactors to the world's energy supplies would be severely constrained. A light-water reactor of 1 000 megawatts electrical output (MWe) operating for 30 years at 70 per cent load factor consumes some 4 000 tonnes of natural uranium over its lifetime. Only about 900 such reactors could be fuelled using the known resources if the spent fuel were not reprocessed to recover and recycle plutonium and residual fissile uranium. While this is a large contribution compared with the 330 000 MWe output capacity of reactors currently projected to be operating in the WOCA region in the year 2000, it is small compared with total electricity generation capacity currently in place, which amounts to 1 593 gigawatts in OECD countries alone (1 GWe = 1 000 MWe).

At current rates of uranium consumption (41 000 tonnes per annum in WOCA), the "known" resources would last about 88 years, which is comparable to the world reserves of fossil fuels (coal, 220 years; oil, 40 years; gas, 60 years) and greater than for many mineral resources.

The nuclear industry has nevertheless endeavoured to reduce concerns about the potential adequacy of future uranium supplies, which became particularly evident in the late 1970s when much higher levels of nuclear capacity (about 1 000 GWe) were widely expected to be constructed for operation by the year 2000. They have done this by producing overall estimates of the quantities of uranium that might exist in "undiscovered resources", largely on the basis of indirect evidence and geological extrapolation. These speculative resources (SR) are estimated to lie in the range of 9.6 to 12.1 million tonnes recoverable at $130/kg U or less in WOCA countries (Table 3), with a further more confidently predicted 0.7 million tonnes in inferred associations with existing deposits and well-defined geological trends (the category called Estimated Additional Resources II, or EAR-II).

Although these resources would take time to locate, delineate and bring into production, they would, if they exist and can be exploited, extend the known resources some threefold. On this basis there appears to be no immediate concern over the adequacy of uranium supplies. Existing mining capacity supported by surplus stocks of uranium built up in the past has exceeded reactor requirements for some years, and will continue to do so until the second half of the 1990s. If the projected nuclear generation capacity growth occurs, additional production centres will be required in WOCA before the year 2000 (Figure 5). This should present no problems, given adequate foresight on the part of producers, and supplies should continue to be provided mainly from low-cost resources.

Two factors could disturb the situation. Any sudden surge in ordering new nuclear generation plants, or even the expectation of such a surge, could increase uranium demand and force prices up in much the same way as was observed during the 1970s, when spot prices rose by over 400 per cent. On the other hand, uranium produced in countries outside WOCA, whose own markets have suffered in the aftermath of Chernobyl, is now being offered in WOCA markets to obtain hard currency. This could

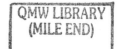
QMW LIBRARY
(MILE END)

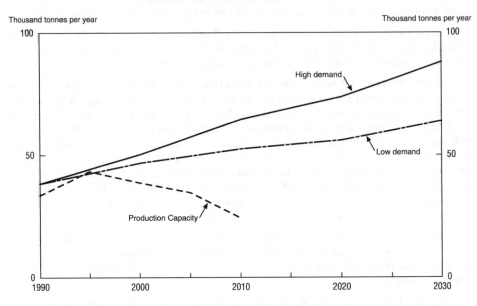

Figure 5. **Projected uranium demand and existing and planned production capacity in OECD**

Thousand tonnes per year

Thousand tonnes per year

Note: Production capacity is for the whole of WOCA.

Source: Uranium Resources, Production and Demand, NEA, 1990; Nuclear Energy Data, NEA, 1991

further depress WOCA markets and force additional mine closures without, in the short term, adversely affecting supplies to OECD countries.

Beyond 2000 there will be a need for a significant increase in supplies, with a growing share likely to come from resources with costs above $80/kg U. On the basis of the capacity projections in Chapter 2 (Table 2), there need be no problem for a considerable period, even if reactor programmes remain largely based on light-water reactors without uranium and plutonium recovery from spent fuel and recycling. However this does presuppose that adequate and timely investment is made in uranium exploration and mining, and that the large bulk of the world's electricity supplies continue to be provided from non-nuclear sources.

The 18 000 terawatt-hours per year (TWh/yr; 1 terawatt = 1 million megawatts) electricity demand projected for WOCA by 2030 in the NEA studies (including 6 000 TWh for developing countries) would consume some 350 000 tonnes of fresh uranium per year if this were the sole fuel in use. This would require most of the WOCA speculative resources as well as the known resources (EAR-I and RAR) to be brought into production early in the next century, and would restrict nuclear fission power to a very limited period of operation before its fuel prices escalated, unless advanced highly fuel-efficient reactors were developed and deployed.

This extreme situation would be exacerbated if, as is quite possible, electricity demand were to grow more rapidly than assumed. Conversely it would be ameliorated if electricity demand were to grow more slowly. In practice, there is no possibility of nuclear power's contribution approaching anything like 100 per cent of total electricity supply in this time-scale, if ever. Nevertheless, the nuclear industry cannot afford to ignore the need to maintain a continuing balance of fuel supply and demand through timely development of uranium resources and uranium-conserving reactor and fuel cycle technologies.

In the long term, uranium supplies from the former USSR, China and East European countries (non-WOCA) appear unlikely to disturb the balance. Their overall energy consumption and total uranium resources (based on IAEA estimates of their speculative resources recoverable at less than $130/kg U using OECD mining technology) are in roughly the same ratio as those of the WOCA region. Assuming parallel development of the two groups of economies, neither region would be expected to be a net importer or exporter of uranium on any scale once the markets have stabilised.

Other fuel cycle services

Lead times of the order of a decade have characterised both reactor planning and construction, and uranium exploration and mine development. They are not shared by the other stages of the fuel cycle, with the possible exception of radioactive waste disposal in deep geological repositories.

Where the need exists, plants for uranium conversion, enrichment, fuel fabrication, new and spent fuel storage, and spent fuel reprocessing and waste conditioning, can all be erected and brought into operation, in countries possessing the necessary technological capability, at a pace matched to the growth of requirements. The only barriers are those of political and public acceptance, often related to specific site selection, particularly where the so-called back end of the fuel cycle, i.e. spent fuel management, is concerned.

It is not surprising therefore that the availability of conversion, enrichment and fuel fabrication services have more than matched demand. Indeed in the late 1980s and early 1990s, market spot prices for conversion and enrichment services have been severely depressed by competition and oversupply which is expected to persist over the next 20 years (Figure 6).

The situation has been complicated by many countries' wishes to develop and make use of indigenous capabilities, even when this has not been the cheapest course, in order to ensure security of fuel supply to their own reactors, independent of world events.

This, combined with the mismatch between projected and eventual demand, for reasons explained in Chapter 2, has led to the current surpluses of capacity. A similar mismatch appears to exist in the former USSR, which has provided fuel for East Europe's Soviet-designed reactors, and which has been providing a growing supply of uranium enrichment services to utilities in the western world (WOCA).

Fuel design, specification and fabrication is a specialised area with fuel companies producing differentiated, high-quality precision-engineered products, carefully matched to the reactor operators' requirements. There is again more than enough capacity to meet current and future needs (Figure 6), with most large countries and several smaller ones having an independent capability.

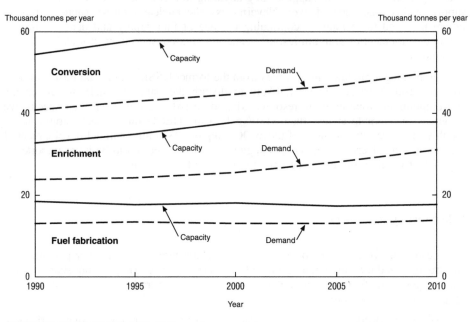

Figure 6. **Annual OECD fuel cycle requirements
and projected production capacity**

Thousand tonnes per year

Thousand tonnes per year

Conversion

Capacity

Demand

Enrichment

Capacity

Demand

Fuel fabrication

Capacity

Demand

Year

Source: Nuclear Energy Data, NEA, 1991

The development of higher burn-up enriched uranium fuels offering economic savings for use in thermal reactors in the United States, Europe and Japan is providing additional opportunities for the fuel fabrication industry to which it is responding as the need arises.

Commercial reprocessing services have only been available up to the present time from France and the United Kingdom. A number of countries have decided not to reprocess civil nuclear fuels for policy and for economic reasons. These include the United States, Canada and Sweden, although the latter had some fuels reprocessed in France in the past. Most European countries have however had, or plan to have, their fuels reprocessed either for reasons of technological necessity (as in the case of some gas-cooled reactor fuels), as part of their waste management policy, or on economic grounds. Japan is now also building a commercial reprocessing plant.

The economic case for reprocessing light-water reactor fuels is significantly dependent on the price of uranium and the costs of enrichment, which determine the value of the recovered uranium and plutonium (see Chapter 5). The depressed state of both of these markets has reduced the anticipated economic incentive for recycling in the short term, although in the longer term it is essential if the full energy content of uranium is to be exploited and nuclear power is to make more than a transient contribution to world energy supplies.

Existing and planned spent fuel reprocessing capacity in France, the United Kingdom and Japan fall far short of the quantities of spent fuel arising across the western world (Figure 7), although a closer balance exists if the North American and Scandinavian countries, which have no plans for reprocessing, are excluded. This is not of great significance since, on present expectations, there is no imperative need to recover uranium and plutonium for recycle in the short term, and spent fuel from water-cooled reactors can be safely stored without corrosion problems for many decades.

Most countries have experienced problems with the identification and acceptance of suitable sites for radioactive waste disposal, particularly for high-level or long-lived wastes and spent fuel. In several OECD countries, local opposition has hampered even the geological research necessary to demonstrate the suitability or otherwise of specific sites. This has not proved critical, since the technology itself is straightforward and the physical need for sites is not urgent and will never be extensive. Nevertheless, lack of real progress in actual implementation of disposal in countries other than Sweden has had a damaging effect on the industry, which is frequently, though unjustly, accused of having no technical means of safely disposing of high-level and intermediate wastes.

The Swedish example could go a long way to dispelling unreasonable public fears and, given the necessary political will, the problem should be resolved during the 1990s. Until this happens and facilities are built, there will be a growing need for supervised surface storage of spent fuel (Figure 7), and wastes arising from reprocessing, from

Figure 7. **Spent fuel management capability in OECD**

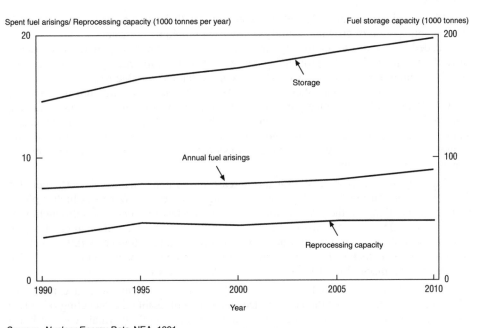

Spent fuel arisings/ Reprocessing capacity (1000 tonnes per year)   Fuel storage capacity (1000 tonnes)

Source: Nuclear Energy Data, NEA, 1991

39

reactor operation and from decommissioning activities which, while safe in itself for several decades, is not an appropriate long-term solution.

Services for advanced fuel cycles

The earlier paragraphs have examined the adequacy of uranium supplies and fuel cycle services on the basis of existing commercial reactor types and conventional fuels. Other fuels that have been developed and are in limited use have the potential to extend greatly the life of the world's uranium resources.

The biggest gains can be achieved by recovering and recycling unused uranium and plutonium from spent thermal reactor fuels. Thus, a 60-fold increase in the energy recoverable from uranium can be achieved using plutonium-fuelled fast breeder reactors that can convert the otherwise useless uranium-238 isotope (99.3 per cent of natural uranium) into useful plutonium fuel.

Plutonium oxide fuels have been extensively used in demonstration and prototype fast reactors and, more recently, in the French commercial-scale Superphénix. These reactors have been supported by suitably sized fuel fabrication and reprocessing plants in the countries developing them.

However, the current expectation that fast reactors will not be deployed on a significant scale until well past the turn of the century has led to an accumulation of stocks of plutonium in spent fuels and separated plutonium, recovered from reprocessed spent thermal reactor fuel, for which no immediate use is planned. This has led to a reconsideration in several countries of the benefits of using mixed plutonium-uranium oxide (MOX) fuels in thermal reactors, and several utilities have been using and plan to expand MOX use.

Mixed-oxide thermal fuels and the plutonium-fuelled fast reactor have been developed in anticipation of growing scarcity of low-cost uranium, a threatened scarcity which has receded in time due to the lower than expected growth of nuclear generating capacity. Nevertheless, the technically proven capability exists to raise the energy recoverable from WOCA uranium resources to a level which exceeds that recoverable from all the world's reserves of coal, oil and gas taken together. The known low-cost uranium resources in WOCA (RAR and EAR-I) are equivalent, when used via plutonium-burning fast reactors, to 4 500 billion tonnes of coal. This compares with some 700 billion tonnes of technically and economically recoverable coal and 200 billion tonnes of coal-equivalent of oil and natural gas. Current total WOCA energy consumption is equivalent to some 8 billion tonnes of coal per annum.

Industry plans indicate a significant increase in the manufacture of MOX fuels for thermal reactors during the 1990s, which will augment their use (albeit at higher plutonium concentrations) in the small number of operating fast reactors. The introduction of 30 per cent MOX fuel into existing light-water reactors is feasible, safe and can become economically attractive as uranium prices rise and MOX fabrication costs fall, if a utility is accumulating a stock of plutonium from the reprocessing of spent uranium fuels. (See Chapter 5 for a discussion of the economics of MOX fuels.)

Overall there are plans to construct MOX fabrication plants with a capacity of 380 tonnes per annum over the coming decade. This will enable the recycling of only a minor part of the plutonium produced annually in the world's thermal reactors but a major part of the plutonium that is planned to be recovered via reprocessing.

Chapter 5

ECONOMIC COMPETITIVITY OF NUCLEAR POWER

The problems of cost comparison

There are strong environmental, strategic and resource conservation arguments for the use of nuclear power. However, the main reasons it has been pursued so vigorously worldwide are the large scale of its potential contribution to world energy supplies and its potential cheapness relative to continuing reliance on fossil fuels.

Although uranium costs around 500 times as much as coal, weight for weight, its complete fissioning yields 2.5 million times as much energy as the complete combustion of an equivalent weight of coal. In a conventional light-water reactor or a gas-cooled reactor, only about 0.5 per cent (0.8 per cent in a heavy-water reactor) of the natural uranium initially required to produce the fuel is fissioned (see Annex note 1), so that the raw fuel costs for a coal-fired plant are around 20 times those for the uranium needed for a thermal fission reactor.

This is an oversimplification, however, since the uranium has to be processed, enriched, and fabricated into engineered fuel elements, which incur further costs for their management after removal from the reactor. Nevertheless, the total fuel costs of producing a unit of electricity using nuclear power are only one-half to one-third of those for a coal-fired plant. Since the fuel costs are the major component of electricity generation costs in coal-fired plants that rely on world-traded coal, this means that electricity produced in nuclear plants will be cheaper than that from coal-fired plants, provided the capital and operations costs of nuclear plants are not disproportionately higher. The much higher energy density of the nuclear fuel provides a smaller volume, relatively compact heat source, and the balance of steam-producing equipment and turbines are essentially common to all thermal power plants whatever their fuel, so that, other things being equal, nuclear plant capital costs would also be lower than those of fossil-fired plants. However, other things are not equal; nuclear plants require special materials for the highly radioactive internal environment they create, they are built with great precision, and they require additional control devices to ensure their safe operation under all foreseeable circumstances.

Their major problem arises from the fact that nuclear fuel, unlike fossil fuels, continues to produce considerable quantities of heat due to the radioactive decay of the fission products it contains, even after the plant has been shut down and the main energy-producing process, fission, has been brought to a halt. This decay heat has to be removed by appropriate physical means if the integrity of the fuel and the nuclear plant is to be maintained. These extra requirements have added considerably to the costs of the specifily nuclear-related section of the generation plant (the nuclear island) and raised total investment costs above those for equivalent fossil-fired plant. Therefore the relative economic attractiveness of the two depends on the balance between the higher capital costs of the nuclear plant and the higher fuel and operating costs of the fossil-fuelled plant.

There are two ways in which costs can be compared. Either by calculating the total incremental costs to the utility of producing all the plants' electricity from one or other fuel; or by allocating the initial capital charges to the units of electricity that will be produced, and adding these to the fuel and operating costs to arrive at separate "total" costs per kilowatt-hour (kWh) for each fuel.

This creates an immediate but not uncommon investment problem, because while the costs of the fuel consumed in both types of plant are almost proportional to the electricity produced, the capital costs are fixed for any given plant, regardless of the electricity it subsequently produces. For purposes of deciding on the better investment, the savings potentially achievable from lower future electricity production costs over an extended period have to be compared with the extra initial capital expenditure.

The problem arises from the fact that an extra expenditure of $10 today is not adequately compensated by a future saving of $10 even in fixed-value dollars (i.e. inflation-corrected). The $10 could have been invested elsewhere and earned real interest (i.e. above inflation) so that more than a $10 saving in the future is needed to compensate for the initial investment expenditure. To overcome this problem and allow for the time-value of money, economists discount future expenditure and income to arrive at their so called present-worth equivalents. (See note 2 of Annex.)

These present-worth costs can be added to produce an equivalent present value of the total lifetime cost of building and operating generating plants which, after standardising to an equivalent electrical output basis, can be used to compare the economic merits of investments.

For the studies undertaken by the Nuclear Energy Agency, the standardised cost derived is the constant-money *levelised lifetime cost* per kWh of electricity. This is an average cost in terms of constant-money value which, if charged for each unit of electricity produced, would exactly repay all the capital, fuel and operating costs, including plant decommissioning, spent fuel management and radioactive waste disposal, and provide a predetermined rate of return on capital (see note 3 of Annex). Comparisons between different generation options can be made on the basis of these levelised costs per kWh.

Results of comparisons

The most recent NEA study, conducted jointly with the International Energy Agency and in association with the International Atomic Energy Agency and the International Union of Producers and Distributors of Electrical Energy, used the above approach to derive the costs of generation from nuclear and coal-fired plants for commissioning in OECD and some non-OECD countries in the latter half of the 1990s. Data on costs were obtained from utilities or national agencies and presented using both individual countries' own economic groundrules and expectations on a common economic basis. The latter retained national investment and fuel costs, but used agreed reference values for plant lifetime, load factors and required rates of return on capital (discount rates), together with variants of these to test the sensitivity of the calculated costs to the underlying assumptions.

Seven out of the ten OECD countries providing data for both nuclear and coal-fired plants found the nuclear electricity option to be significantly cheaper (i.e. more than 10 per cent cheaper) using their own assumptions, while the remaining three showed broadly equal costs (Table 4). Of the non-OECD countries, four showed a significant nuclear advantage, while four showed approximate parity and one (Hungary) showed coal to be significantly cheaper. The above comparison excludes the cheap-coal regions of the United States, Canada, Brazil and China, where coal plants situated adjacent to their fuel source would be the cheaper option.

The common reference assumptions of 30-year plant life and 72 per cent load factor were both amply supported by operational experience for both coal and nuclear plants. A 5 per cent real return on capital was taken as the reference value in line with that employed by 11 of the 17 OECD countries reporting the parameter. Using these values, the comparative position was the same for OECD (Table 4; the same seven countries showed nuclear to be cheaper), but now all of the non-OECD countries except Hungary and Czechoslovakia also showed nuclear electricity to be significantly cheaper than that from coal. This was due to the higher required rates of return (typically 10 per cent per annum) generally used in the non-OECD countries' own calculations. (Data for Hungary and Czechoslovakia were obtained by IAEA after completion of the NEA study).

Table 4. **Comparison of nuclear and coal-fired generation costs
for plants commissioned in 2000**

| Country | Ratio of coal generation to nuclear generation costs | | |
| | Common assumptions | | National assumptions |
	5% ROR*	10% ROR	
Belgium	1.79	1.42	1.35
Canada (Central)	1.33	1.06	1.41
Finland	1.20	0.98	1.21
France	1.45	1.22	1.32
Germany	1.42	1.13	1.24
Italy	1.07/1.43	0.86/1.08	1.02/1.34
Japan	1.28	1.11	1.17
Netherlands	0.95	0.80	0.99
Spain	0.97	0.77	0.97
Turkey	1.05	0.79	–
United Kingdom	1.06	0.85	0.96
United States (East)	1.07	–	–

* ROR: Rate of Return.
Note: Data for Germany are for domestic coal; cheap-coal regions of North America are omitted.
Source: *Projected Costs of Generating Electricity,* NEA, 1989.

Although, contrary to public perceptions, the projected costs of nuclear electricity have remained fairly steady, taken overall, over the period 1983-1989 covered by successive NEA studies, nuclear power has lost ground to coal due to the sharp decline in coal prices that occurred in the mid-1980s and its effect on the utilities' projected levels of future coal prices. This is despite the somewhat increased capital and operating costs now projected for coal plants in those countries that had not in the earlier studies planned to introduce flue-gas desulphurisation technology.

Obviously if coal prices were to prove lower than expected by the utilities or if higher rates of return on capital were sought, coal would become more competitive. The sensitivity to discount rates for a range of countries is shown in Table 4. At a 10 per cent discount rate, the highest employed in any of the OECD and non-OECD countries reporting data, only four OECD countries continued to show a significant (about 10 per cent) advantage to nuclear power, three show comparability and four a significant advantage to coal. Of the non-OECD countries, two move from showing a nuclear advantage to near break-even compared with the position based on their own assumptions.

It is clear that in most regions of the world except those with direct access to low-cost coal, countries have expected nuclear power to remain cheaper or at worst break even in cost terms with coal-fired plant. This nuclear advantage is not such, however, that it could not be removed if coal prices were to remain significantly below those expected by utilities, or if significantly higher real rates of return on capital were to be required. (The rates of return at which coal and nuclear power break even are shown in Table 5). On the other hand, higher coal prices arising either from market forces or the imposition of environmentally orientated carbon taxes, or any general reduction in required rates of return on capital could markedly increase the perceived nuclear advantage.

Each utility or country will have its own views on the future course of events and its own views on nuclear versus fossil fuel competivity. These views will evolve in the light of changing national and world economic and political development. It is already apparent that natural gas-fuelled combined-cycle power generation, which was considered uneconomic in the 1989 study, is now viewed more favourably in several OECD coun-

Table 5. **Break-even rates of return**

Plant lifetime thirty years

Country	Rate of return
Belgium	17.0
Canada (Central)	12.0
Finland	10.6
France	17.6
Germany	12.2
Italy	6.6-11.2
Japan	13.6
Netherlands	3.6
Spain	4.5
United Kingdom	6.0
United States	6.4

Source: Projected Costs of Generating Electricity, NEA, 1989.

tries. An update of the earlier studies is in progress and will reflect any changes in expectations that have taken place during the past few years.

The structure of nuclear generation costs

a)   *Background*

Nuclear (or fossil-fired) generation costs are conventionally split into three main components: capital investment, fuel, and non-fuel operation and maintenance costs. There are more minor components such as plant decommissioning and ultimate dismantling costs, plant-related R&D, or post-operational capital expenditures.

There are also differences in the way in which utilities choose to categorise costs and treat them in investment appraisal. For example, the costs of the initial fuel charge for a nuclear plant and of plant spares may be capitalised as part of the initial investment or treated as part of the fuel costs and operations costs respectively. Such differences are of no consequence, provided a standardised approach is used for comparisons between countries (as it has in NEA studies) and all the costs are properly captured in the analysis.

The costs of nuclear electricity are clearly sensitive to the costs of the separate main components, and NEA studies have thrown more light on these. They are discussed separately below.

b)   *Investment costs*

Capital (or investment) costs are the major contributor to the overall costs of nuclear electricity generation (Figure 8). If investment costs are examined in more detail, it is evident that there are considerable differences in "overnight" investment cost expectations among countries (i.e. the direct investment costs excluding any interest charges incurred during plant construction). This divergence (Table 6) has been apparent from the earliest studies and arises in part from the differences in the costs of land, labour, materials, etc. (factor costs), in regulatory approaches, siting requirements, design choices and from exchange rate anomalies. There is however clear evidence that countries replicating designs and planning the construction of a series of plants on a single site do achieve significant economies that improve nuclear power's overall competitive position. France and Canada in particular have benefited from coherent strategies of this type.

The overnight investment cost is not the only factor of importance however. The capital tied up unproductively from the start of construction until a plant is connected to the electricity grid and commissioned, costs the utility money, normally treated as interest on the capital employed during construction (IDC). This interest increases as the time taken to build and commission the plant increases, and it can become a considerable part of the overall investment cost (Table 6). Depending on the phasing of expenditure and construction time, IDC can range from as little as 14 per cent of investment cost for plants built in five years to as much as 30 per cent or more for plants taking twice as long (with interest rates of 5 per cent per annum).

Countries differ considerably in their past experience of nuclear plant construction. Some achieved particularly good results while others have encountered licensing

Figure 8. Composition of levelised discounted nuclear electricity generation costs
(30 years lifetime, 5% discount rate)

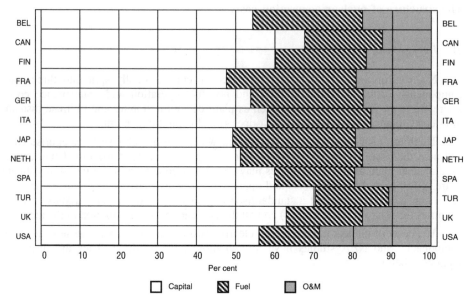

Capital ☐ Fuel ▨ O&M ▦

Source: *Means to Reduce the Capital Cost of Nuclear Power Stations*, NEA, 1990

Table 6. Capital costs, interest payments and decommissioning provisions

Country	January 1987 US$/kWe		
	"Overnight" capital cost	Interest during construction[a]	Decommissioning provision[b]
Belgium	1 327	211	29
(Central)	1 138	169	12
Finland	1 387	217	28
France	1 093	171	34
Germany	1 879	275	35
Italy	1 925	378	44
Japan	1 773	252	38
Netherlands	1 414	219	62
Spain	2 060	374	61
United Kingdom	1 828	420	28
United States (Midwest)	1 406	296	100

a) Based on 5% per annum rate of return.
b) Based on country assumptions. U.S. figure is undiscounted.

Source: Projected Costs of Generating Electricity, NEA, 1989.

problems or introduced late design changes that have led to protracted construction periods and high interest costs.

c) Nuclear fuel costs

Nuclear fuel costs remain a significant but minor part of the total nuclear generation costs. Separate studies have examined their make-up in detail and reflected on likely future trends. For light-water reactor fuel, the feed uranium costs and enrichment costs are the major front-end (pre-reactor) components (Table 7). Uranium prices, which were widely expected to rise as nuclear capacity expanded, have in practice dropped enormously since the 1970s as a result of the failure of demand to grow at the levels widely predicted and which were, in consequence, catered for by the mining companies. Uranium stockpiles built up by suppliers and utilities in anticipation of growing demand have been returned to the market, further depressing new requirement levels and prices (see Chapter 4).

Developments in enrichment technology have also reduced its energy requirements and brought the costs of enriched uranium down, so that prices of fabricated enriched uranium oxide fuel have been significantly lower than would have been projected in the 1970s.

Table 7. **PWR fuel cycle component costs**

US mills of January 1984

	Reprocessing cycle (percentage)	Once-through cycle (percentage)
Uranium	26.5	35.8
Conversion	2.2	2.5
Enrichment	29.6	34.0
Fabrication	11.4	12.9
Sub-total for front end	74.5	85.5
Transport of spent fuel	1.8	2.1
Storage of spent fuel	2.2	9.7
Reprocessing/waste vitrification	28.3	–
Spent Fuel conditioning/disposal	–	2.7
Waste disposal	1.0	–
Sub-total for back end	33.4	14.5
Uranium credit	–7.9	–
Plutonium credit		
Total fuel cost	100	100
Total fuel cost (mills/kWh)	7.7	6.7

Note: Based on uranium at $32/lb U_3O_8 ($83.2/kg U).
Source: The Economics of the Nuclear Fuel Cycle, NEA, 1985.

The costs of the back-end of the nuclear fuel cycle, i.e. management of spent fuel after its removal from the reactor, still have elements of uncertainty. Apart from the commercially available reprocessing services, there is no established market on which to base prices for spent fuel storage or for spent fuel or reprocessing waste disposal. The costs of storage or waste disposal are sensitive to individual countries' regulatory requirements and to the scale of their operations. Work is continuing to develop an international consensus and many studies have been made of the technologies needed and their costs.

However, the broad magnitude of their costs is clear, and the back-end uncertainties amount to 10 per cent at most of the overall fuel cost, or a few per cent of total nuclear generation costs. It is also evident that the overall costs of the nuclear fuel cycle for thermal reactors will remain at levels similar to those of today, with limited real increases as and when uranium prices rise in the post-2000 period.

The economic attractiveness of reprocessing spent fuel rather than subjecting it to indefinite storage (and possible direct disposal to geological repositories) is dependent on both storage and reprocessing costs, and on the costs of uranium and of its enrichment. If the latter front-end costs are low, then the value of recovered plutonium and unburnt uranium-235 for recycling in thermal reactors is also low, and the economic incentive for their recovery via reprocessing will be small or non-existent in the short term (see note 4 of Annex).

It is unlikely that the construction of new reprocessing plants could be justified on economic grounds alone at the present time, but the differences in cost per kWh between the spent fuel storage and disposal (once-through) option and the reprocessing option is not large (Table 7). Waste management and disposal considerations, the wish for ultimate strategic independence from imported nuclear fuel supplies, and the wish to develop and demonstrate the relevant fuel technologies, for example, may be a sufficient justification for incurring some extra short-term costs in pursuit of expected longer-term benefits in some countries. Analysis of the overall levelised costs of uranium oxide fuel over the lifetime of a thermal reactor has shown these to be insensitive to changes in parameters like reactor life and the required rate of return on capital.

Fuel costs can be reduced, however, by extending fuel burn-up, i.e. increasing the energy recovered per kg of fuel. This reduces spent fuel management charges per unit of power obtained and can assist in improving the availability of the reactor itself by reducing the frequency of, or the time taken for, refuelling operations. The optimum burn-up is a function of uranium and enrichment costs, since higher uranium-235 enrichments are required to make high-burn-up fuel.

d) Non-fuel operating costs

The other component of generation costs is the non-fuel operating cost of the plant. Most countries and utilities have found this to be a relatively small cost, which is comparable to that incurred in running a coal-fired plant, despite the considerable differences in the activities involved. Some expect future nuclear plant operating costs to be lower and some higher than those for coal plants.

The major exception is the United States, where nuclear plant operating costs soared during the 1980s to become more than twice as large as the nuclear fuel costs for the reactors (Figure 8). This escalation for existing plants has been due in part to the specific regulatory problems US operators have experienced and to the programmes of retrofitting to meet changing safety requirements following the Three Mile Island accident. However, the magnitude of the projected divergence between the United States and other countries in the NEA study for future new plants has not been fully explained, and may have arisen from differences of definition. Clearly it would be detrimental to nuclear power's competitiveness if the costs of running the plant were to increase significantly above the general level of international experience. The current NEA updating of generation costs will provide fresh guidance on the trends in operating costs being experienced in OECD countries.

e) *Decommissioning*

Decommissioning costs for large 1 000 MWe light-water reactors are high in absolute terms, in excess of $100 million, although this is a small proportion (10 per cent to 20 per cent) of the initial capital cost of the plant and a very minor component of overall electricity generation costs.

Reactor decommissioning, like radioactive waste disposal, remains a matter of public concern despite the industry's confidence that it has safe and well-understood techniques for dealing with both problems. These public concerns may well persist until a number of specific successful demonstrations have been performed, despite the fact that several international and national studies have amply demonstrated that the technical methods and equipment are already available to accomplish the dismantling of nuclear reactors and other nuclear facilities safely, whatever their type and size.

The industry's confidence is based on the experience that has already been gained through the use of the same techniques in the course of maintenance and repair work on reactors and plants, and in the decommissioning of earlier demonstration, prototype and small power reactors. Additionally, a number of specific decommissioning projects are under way involving commercial-scale facilities. These will give added confidence in the technologies well before the bulk of today's commercial reactors reach the end of their economic lives.

At the present time, estimates of the costs of decommissioning commercial reactors and nuclear fuel cycle facilities are based upon the worldwide experience that has been gained in the smaller though similar tasks of decontamination, dismantling and waste disposal that have been undertaken.

There remains a significant divergence of views between countries over the precise costs of reactor decommissioning (Table 8). Attempts to establish the reasons underlying these divergences have revealed that they are similar to those affecting the projected costs of reactors in different OECD countries. Differences in reactor design, differences in the regulatory framework, differences in factor costs and differences in plant size all play a significant role. Another major factor in comparisons between country estimates is the effect of currency exchange rates, which are largely determined by factors other than the physical resource requirements (including manpower and materials) associated with industrial operations.

In the case of nuclear plant decommissioning, the specific policies of individual countries or utilities with regard to the timing of the decommissioning activities can have a significant impact. Whereas most countries would begin to remove the remaining spent

Table 8. Decommissioning costs for LWRs
January 1991 US$

Country	Reactor type	Cost per unit capacity $/kWe	Cost per unit of waste $/kg
Finland	PWR	250	30.8
	BWR	130	17.7
Germany	PWR	175	20.5
Japan	PWR	190	19.1
	BWR	210	20.7
Switzerland	PWR	140	32.0
	BWR	190	27.4
United Kingdom	PWR	360	33.4
United States	PWR	100	14.1

Notes: Based on reactor gross capacity and untreated waste volumes. Costs are undiscounted.
Source: Decommissioning of Nuclear Facilities, NEA, 1991.

fuel from the reactor core immediately following final reactor closure, the timing of the subsequent dismantling operations, both for the non-nuclear associated facilities and for the reactor's pressure vessel and core itself, are matters of choice. Because the reactor's pressure vessel and core are highly radioactive, there are benefits in deferring the final stages of dismantling in order to allow as much as possible of this radioactivity to decay before operations commence. This both reduces radiation exposure during operations and permits the use of simpler less costly techniques. (Chapter 3)

In economic terms, the delay in stage-2 and stage-3 decommissioning also reduces the contribution of the decommissioning costs to overall generation costs. In simple terms, the money that would need to be set aside to provide funds for ultimate reactor decommissioning will be smaller the longer the time those funds are available to earn interest, since the ultimate decommissioning cost, whatever it may be, is fixed in constant-value money terms.

Table 8 sets out the decommissioning estimates produced in recent studies. Based on these values, it can be concluded with confidence that the cost of decommissioning will contribute no more than a few per cent to the overall levelised costs of electricity production in light-water reactors, despite the wide range of costs projected in different countries.

The costs of dismantling some gas-cooled graphite-moderated reactors is projected to be significantly higher than that for light-water reactors due to the larger volume of wastes associated with the former reactor type, particularly in the case of the earlier low-power-density reactors (Magnox) using metallic uranium fuels. When standardised in terms of decommissioning cost per unit of decommissioning waste, the wide divergence of costs between gas-cooled and water-cooled reactors and between different countries' water-cooled reactors is greatly reduced, though not eliminated, for the reasons set out earlier.

The economics of plutonium fuels

There is a small but growing use of plutonium-uranium mixed-oxide (MOX) fuels in commercial light-water reactors (see Chapter 3). Their use in thermal reactors is based not on need, as in the case of the fast reactor, but on the perception that it will save utilities money, if not immediately, then in the near future.

Recent NEA studies on the economics of mixed-oxide fuels re-examined the costs of the separate stages of uranium fuel production and confirmed that they are, if anything, lower than those projected in the detailed studies of uranium fuels conducted in 1985 (which are currently being updated). The mixed uranium-plutonium oxide fuels themselves have been examined in terms of the costs of providing the annual fresh fuel reload (normally one-third of the core fuel is replaced) in an existing light-water reactor that has previously been using uranium fuels. This is the most likely application in the 1990s and simplifies the cost comparisons.

The study identified the circumstances under which the use of fuels containing plutonium would be economically attractive to utilities. There are no inherent operational or safety problems for most existing reactors, provided the percentage of mixed-oxide fuel in the core is limited to about 30 per cent. For utilities with separated plutonium stocks, or contractual obligations leading to such stocks, their use in plutonium fuels will save money in the future, provided the fabrication costs of the mixed-oxide fuels can be brought down through the use of larger fabrication plants.

The mixed-oxide fuels (Figure 9) avoid the need for costly enrichment by substituting plutonium for uranium-235. They can also avoid the need for any fresh uranium through the use of stocks of uranium derived from spent fuel reprocessing or of depleted uranium from enrichment plant tails. Unenriched natural uranium can also be used, but the quantities needed are small relative to those used to produce enriched uranium. In each case, blending of plutonium and uranium oxides can produce a fuel with similar fission characteristics to the uranium oxide fuel for which it substitutes. The cost savings are however offset by the higher costs of mixed-oxide fuel fabrication. These higher costs are a consequence of the additional containment measures needed to accommodate the higher radiotoxicity of plutonium and the penetrating gamma radiation emitted by trace impurities in the recovered fuel. Given the extra costs of fabricating one kg of mixed-oxide fuel relative to the costs of fabricating one kg of enriched uranium oxide fuel, it is possible to calculate values for the uranium and plutonium recovered by reprocessing, based on the cost of fresh natural uranium and enrichment services. These values are the substitution values of recovered uranium and plutonium associated with their use in light-water reactors. They would be different for different reactors and for different levels of fuel enrichment.

It will become economically worthwhile to reprocess spent uranium oxide fuel when the value of the recovered uranium and plutonium is equal to or greater than the net costs of reprocessing. These net costs are determined by the difference between the unit costs of reprocessing and the unit costs of prolonged storage of spent fuel (see note 4 of Annex).

For MOX fuel to be economic, it is clearly necessary that its extra fabrication costs be smaller than the costs of natural uranium and enrichment services required to produce uranium oxide fuel. Figure 10 illustrates the cost comparison per kWh of electricity based on a uranium price of $80/kg U, which is high relative to current spot market prices.

Figure 9. **The stages of the uranium oxide and mixed-oxide
fuel cycles for the PWR**

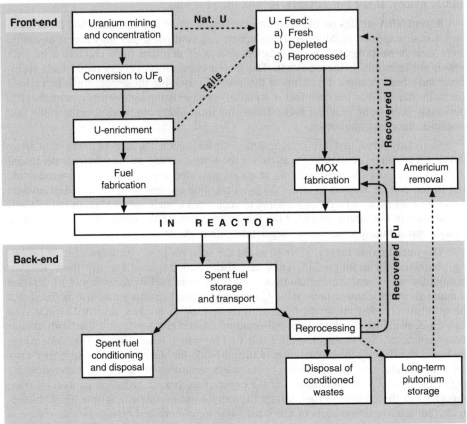

Figure 9. **The stages of the uranium oxide and mixed-oxide
fuel cycles for the PWR**

Uranium oxide fuel   MOX fuel

Source: *Plutonium Fuels : An Assessment,* NEA, 1989

Account will be taken, however, of any savings that could be made through the avoidance
of the need to store plutonium that has already been separated through reprocessing.
When the use of separated plutonium is economically attractive at given prevailing
market prices, there remains the question of whether it would be economic to pay to have
it recovered from spent fuel in an existing reprocessing plant, or whether it would pay to
construct a new reprocessing plant specifically for the purpose of recovering the plutonium for recycling in thermal reactors.

At current uranium market and enrichment prices, it would certainly not pay economically to build a new reprocessing plant. Where plants already exist however, and their costs have been wholly or partly paid, the situation is more favourable.

The actual savings achievable from plutonium recycle will be enhanced for higher burn-up fuels that call for higher fissile material content (Figure 10). There are also advantages in the prompt reuse of plutonium after its separation, since this reduces the build-up of gamma-emitting americium-241 produced by the radioactive decay of plutonium-241. This gamma radiation complicates handling, and the plutonium decay marginally reduces the fissile energy value of the fuel. The co-reprocessing of spent MOX and uranium oxide fuels will also improve the overall economics of MOX use, since it reduces the unit cost of plutonium recovery, though this is not likely to be a significant factor in the 1990s.

Repository costs for storage and disposal of spent fuel are site and scale-dependent, and different countries have different radioactive waste management strategies, so that it is not altogether surprising that there is a divergence of view on the value of reprocessing and plutonium recycle.

Figure 10. **Illustrative costs per kWh for equilibrium replacement PWR fuels (using free plutonium)**

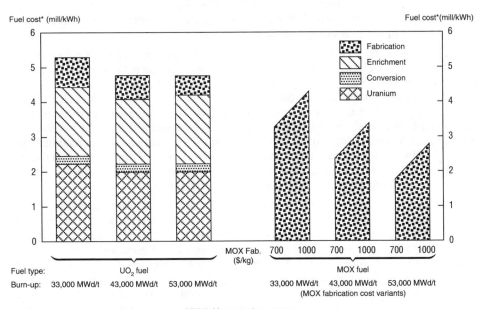

MWd/t: Megawatt-days per tonne

* Discounted cost at the charge of reactor (discount rate: 5%)

Source: Plutonium Fuels: An Assesment, NEA, 1989

Several European countries and Japan are either already using or actively planning to use MOX fuel in their light-water reactors, perceiving this to be an economically attractive option in their circumstances. Others like the USA and Canada have no such plans because they see no economic advantage in the foreseeable future. The United Kingdom is a reprocessing country which plans to produce MOX fuel, but it has no current plans for its indigenous use in thermal reactors since its economic attractiveness is smaller in the UK's existing gas-cooled reactors.

Comparison with other electricity generation technologies

There have been few international comparisons of the generation costs from other fuels or technologies. Many cost estimates exist in the literature, particularly for renewable sources, but these are not based on well-authenticated data produced by experienced utilities using the same basic methodology and assumptions as those employed in the nuclear and coal generation cost studies described earlier in this chapter.

One reason for this is that both oil and gas fuels appeared too expensive relative to coal to compete for baseload power generation in most OECD countries during the 1980s. The data supplied to the NEA/IEA in the course of their 1989 study showed that utilities doing studies expected this situation to persist through the 1990s. This contrasts with the more recent view in the United Kingdom in particular, where gas has become available on long-term contract at low prices (by international large-scale supply standards), that gas-fired combined-cycle plants will be a cheaper source of baseload electricity in the near future than either coal or nuclear power.

It remains to be seen how much gas can be made available at these prices in the United Kingdom or elsewhere, particularly in view of the strong environmental incentive to switch from coal to gas for many industrial and domestic heating purposes, as well as for power production, because of the lower emissions of carbon dioxide and acid gas combustion products that gas produces. However, there is now a general expectation that gas's contribution to electricity supplies will increase significantly in OECD countries over the next decade, assisted by the significantly smaller capital investment requirements and the relatively rapid construction periods anticipated; both of which are attractive to electricity utilities. It is expected that this changed perception of gas prospects will be reflected in the current updating of the generation cost studies.

The economics of renewable sources are inevitably highly variable, because they depend significantly on geographical and climatic circumstances. Hydropower, tidal power and geothermal power are all very site-dependent. Only wind turbines and run-of-river hydrogenerators can be regarded as transferable technologies, and even their performance and unit generation costs will depend on local conditions. On the basis of data provided to the NEA, wind turbines were not expected to be able to compete with fossil-fuelled or nuclear plants in the absence of special incentives. They may, however, prove attractive in isolated regions for fuel-saving purposes where the use of large-scale conventional plants is inappropriate, and their costs could decline if they are produced in significant numbers.

No participating utilities provided cost data on solar power, biomass, waste incineration, wave energy or advanced coal-burning technologies, due in part to the relatively early stage of their technical development, and in part to the absence of reliable cost and performance experience.

This is not to say that these technologies will not be utilised during the coming decade, but the extent of their use will be at best limited and exploratory. Even if they prove successful in some circumstances, their contribution to overall electricity supply will be constrained by environmental and logistic difficulties in the foreseeable future.

Overall it is likely that gas-fired plants and renewables, where they prove economic, will displace large coal-burning plants rather than nuclear plants, for environmental and cost reasons. This could however have an impact on the ultimate penetration of nuclear power into the world's electricity generation systems.

Chapter 6

GOOD PERFORMANCE AND FUTURE DEVELOPMENTS

Reactor development

a) General considerations

As noted in Chapter 3, the early years of nuclear power were marked by the parallel development of a number of different reactor concepts and designs. Over the years these developments have come to focus on five main reactor types: the pressurised light-water-cooled reactor, the light-water-cooled boiling-water reactor, the pressurised heavy-water-cooled and moderated reactor, the liquid-metal-cooled fast reactor and the gas-cooled high-temperature reactor. Of these only the first three are deployed in significant numbers commercially.

During the period of development, there have been pressures on the already safety-conscious industry to aim for ever higher standards of safety assurance, backed by detailed analysis of all conceivable modes of failure arising from single events or sequences of events, no matter how improbable.

The evolving requirements inexorably led to more complex reactor designs with many replicated safety systems, and this contributed to an escalation in real costs that has weakened nuclear power's competitive position in many countries; a position which has been made even more difficult by the sharp decline in fossil fuel prices from the mid-1980s onwards. A further problem that has added to nuclear generation costs has been the comparatively poor availability performance, relative to initial expectations, of nuclear plants in a number of OECD countries. If a reactor achieves an availability of only 60 per cent rather than a reasonable target of 80 per cent, the effect is equivalent to an increase in the capital cost component of a unit of electricity of 25 per cent.

In these circumstances, it is scarcely surprising that the nuclear industry has attached highest priority to getting good performance from its existing plants and to finding means of reducing the capital costs of new plants, while not in any way reducing their safety.

b) Improved reactor performance

A first priority for any nuclear utility is to get the best possible performance out of the nuclear plant it has available. A great deal of experience has been accumulated in OECD countries and a number of common approaches have emerged from which all countries can benefit.

In some cases where spare turbine capacity has existed, it has proved possible to upgrade a reactor to obtain significant additional power for a relatively modest investment in improved safety systems. The opportunity is not available for the majority of reactors and may prove a more practicable option when repeat orders are being placed for new plants, since these can allow marginal adjustments to be made to plant design, without significantly affecting its costs or necessitating major new safety reviews.

Widespread exchange of information and analysis of the frequency, causes and duration of unplanned shut-downs is serving to highlight the common problem areas, and offers the opportunity of identifying factors that contribute to good or bad performance. Comparison of experiences and technologies can lead to a general improvement in plant management and reactor availability.

It is recognised that good management is essential from the top levels down. It ensures that sufficient well-trained and qualified staff are available to carry out all routine operations. These operations should be meticulously planned and experience recorded so that means of further improvement can be identified.

For example it is now routine practice for refuelling, maintenance, component replacement and safety inspections to be dovetailed together to reduce planned outage times (shut-down periods) to a minimum. Wherever possible, use will be made of

Figure 11.   **On-line scrams for French 900 MW and 1300 MW units**

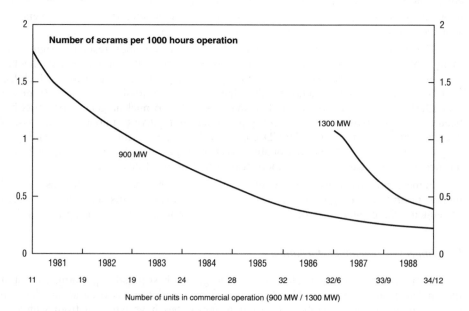

Source: Good Performance in Nuclear Projects, NEA, 1989

58

automated remotely controlled equipment to speed component testing and to reduce the exposures of maintenance staff to radiation.

Unplanned outages have been reduced by replicating components found to give rise to problems, giving preference to changes that permit the replacement of failed items without a need to shut down the plant. Planned preventive maintenance has been adopted, with the regular replacement of selected components at frequencies calculated to eliminate failures during plant operation. The operational service life and reliability of individual components have been improved by modifying their design, and by implementing effective quality assurance programmes to ensure that all items are fully tested before they are put into service.

The frequency of unplanned outages may, in some cases, also be reduced by improving reactor instrumentation and control systems, so that fewer unnecessary shutdowns are occasioned as a result of the need to allow excessive safety margins to cater for the imprecision of existing systems.

Given management commitment, with good planning and training and constant vigilance, it has proved possible to improve greatly the performance of reactors in the majority of OECD countries. Considerable reductions have been achieved (Figure 11 shows French experience) in the frequency and duration of unplanned shut-downs (scrams). The duration of planned outages has been decreased and the radiation exposure

Figure 12. **Mean refuelling shutdown times for German KWU PWRs**

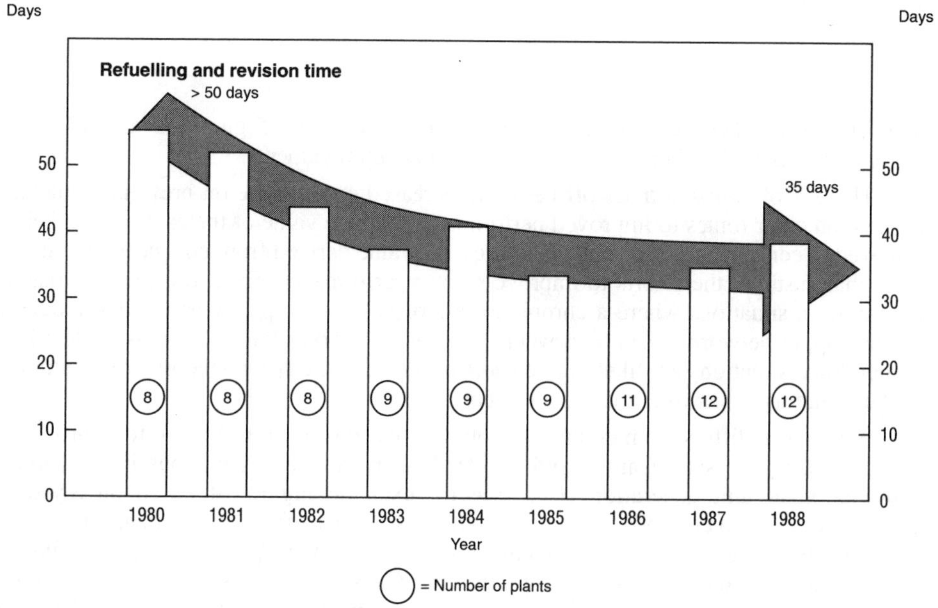

Source: *Good Performance in Nuclear Projects,* NEA, 1989

59

Figure 13. **Personnel exposures in 1988 in Germany**

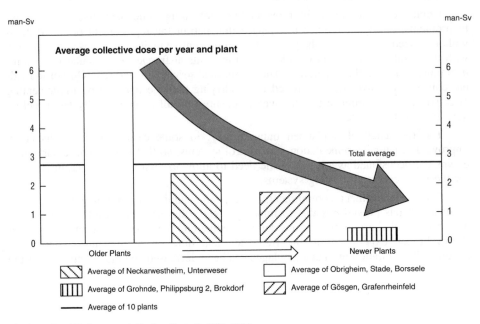

Source: *Good Performance in Nuclear Projects,* NEA, 1989

of operators has been considerably reduced over the years. Figures 12 and 13 show German experience. Other countries show similar improvements.

There is of course a trade-off between increased expenditure on preventive mainte-nance (and other routes to improved performance) and the value of the extra output being achieved in consequence. For most countries, the value derived from enhanced output has more than justified the efforts to improve reactor performance. Exceptions could arise in the future in situations where a chronic excess of baseload capacity above that required by the supply network exists, and where there is no market for the increased electrical output. Such situations are likely to be short-lived if electricity demand growth continues in the future as it has over the past 40 years.

Experience differs from country to country, but most have been able to improve on initial performance significantly, and this leads to worthwhile reductions in the capital contribution to unit generation costs (see for example Figure 14 based on Swedish experience). However, several countries have also had to face unanticipated technical problems that have retarded or even temporarily reversed their progress. The cracking of pressure tubes in the normally high-performing CANDU reactors is but one example. Such problems can be overcome and will reduce in number as experience accumulates, so that the general trend of improvement will be maintained.

Figure 14. **Energy availability for Swedish BWRs**

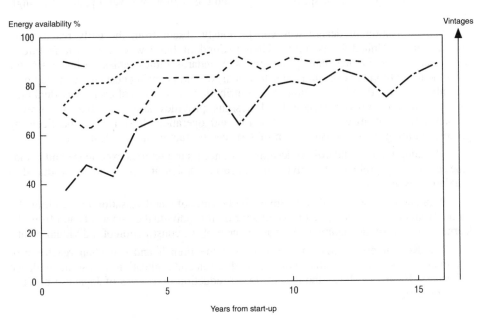

Source: *Good Performance in Nuclear Projects,* NEA, 1989

c) *Capital cost reduction*

A second priority for reactor operators is the reduction of the capital cost of new plants, to the extent that this can be achieved without adversely affecting safety standards. In the past, one route to capital cost reduction has been perceived to be a progressive increase in the size of individual plants which, in principle, should allow reductions in their specific costs (i.e. costs per kW of electrical capacity). Individual reactors have grown over the years from a few hundred MWe of capacity to nearly 1 500 MWe; sizes that would have seemed inconceivable to the power engineers of the 1940s.

The size of individual units is limited partly by materials technology and partly by the scale of the contribution that individual plants can be allowed to make to the grid system they are serving. Concentrating power supply in a small number of very large units, even when they have very high availability factors, can adversely affect the overall stability and reliability of a network.

While countries like France, Japan, Germany and the United Kingdom have large power grids and have been firmly committed to large plants, there has been a growing debate over the advantages of smaller reactors; advantages that may offset the benefits of scale that such reactors forego. This debate is explored further in the section describing small reactor development.

The main routes to achieving capital cost reductions appear to be by a standardisation of design, replication of specific designs and collocation of several plants at a single site.

Standardisation of design was conspicuously absent from the early reactor programmes in the United States and the United Kingdom, but it was an intrinsic feature of the Canadian and French main programmes. Standardisation enables design costs, the manufacturing tooling costs, and the safety analysis and licensing costs to be spread over a number of units, thereby reducing their contribution to the costs of each individual unit. The adoption of proven standardised designs provides assurance that unforeseen problems will not arise during construction and operation. It simplifies the licensing process, limiting it to a consideration of site-specific factors in the ideal case.

The adoption of standardised designs gives access to previous experience and avoids repeated teething problems by relying on tested equipment and components produced to standard specifications.

Replication of specific designs gives the benefits of standardisation but additionally profits from the increased scale of manufacturing for individual components and from the learning process in the planning and management of the construction of individual plants.

The Konvoi programme in Germany and the French and Canadian reactor programmes have sought to capitalise on standardisation and replication with some evidence of success. Figure 15 shows the expected cost reduction in a series of Japanese boiling-water reactors.

Figure 15. **Cost reduction for Japanese BWRs, in constant 1990 Yen**

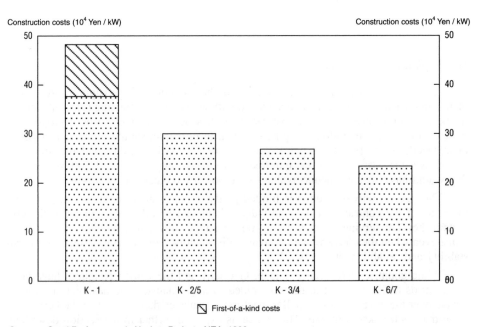

Source: Good Performance in Nuclear Projects, NEA, 1989

d) Collocation

A number of significant advantages have been gained by the planned and co-ordinated siting of a number of reactors at a single site. The purchase of the site and the costs of the transmission lines to connect the site to the central power grid do not increase in proportion to the capacity installed at the site. With good forward planning, only one planning application needs to be made for the site and its transmission lines, with a consequent saving in the time and the costs of obtaining the pre-construction approval, in accordance with the regulatory and planning requirements of the individual country.

Given a planned continuous programme, it is possible to maintain continuity of employment and create a pool of experienced and skilled workers, which facilitates the construction phases. Continuity of employment can contribute to increased labour productivity by removing the disincentive effect of threatened redundancy when a single job is completed.

The collocated reactors are able to benefit from common services and facilities such as access roads or railways, water supplies, administrative and other staff-related functions.

Finally the location of a number of reactors on one site leads to technical savings in relation to storage facilities for fresh or spent fuel. It reduces the size of the stock of spare components that has to be held relative to that needed for separately sited plants. It can reduce the total requirement for backup safety and maintenance services, and it can make more effective use of the specialised skills of operational staff once the plants are in full operation.

Table 9. **Estimated cost reduction between the first and second units of a two-unit station in the U.S.A.**

Unit size of 1 200 MWe each

Account	Unit 2 cost / Unit 1 cost (%)
Total structures and improvements	79
Total reactor plant equipment	91
Total turbine plant equipment	95
Total electric plant equipment	83
Total miscellaneous plant equipment	60
Total main condenser heat rejection system	87
Total direct cost	86
Total construction services	57
Total home office engineering and services	35
Total field office engineering and services	64
Total indirect cost	53
Total base construction cost	69

Source: *Means to Reduce the Capital Cost of Nuclear Power Stations*, NEA, 1990.

There is evidence from several countries, including Canada, France and Japan, for the cost reductions associated with second and subsequent plants when they are constructed as part of coherent programme on a single site. Cost reductions amounting to some 20 per cent or more have been achieved by this means (Table 9).

e) Reduced construction time

Interest charges on the capital employed during the construction of nuclear plants can become a considerable proportion of the total cost (Chapter 5, Table 6). There has been a wide variation in experience of reactor construction both between and within countries, and a great deal of attention has been given to methods of ensuring that construction proceeds smoothly and efficiently to the earliest possible completion.

As in the case of reactor operation and maintenance, good management and meticulous planning and scheduling are an essential prerequisite. Every aspect of the construction process needs to be thought through carefully using the latest techniques of computer-aided planning, supported by the construction of scale models in order to anticipate problems before they have to be faced in practice.

Apart from the application of the now standard techniques of critical-path analysis to ensure that all materials, components and labour are available at the right place and at the right time, attention has to be given to the possibility of using prefabrication techniques to

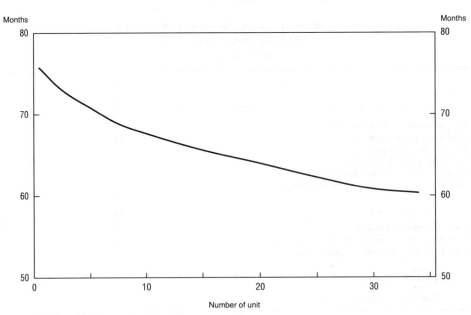

Figure 16. **Trend of French 900 MW PWR construction times, from boiler commitment to grid connection**

Source: Good Performance in Nuclear Projects, NEA, 1989

64

reduce the volume of on-site work and minimise susceptibility to disruption arising from adverse weather conditions. Wherever possible, automated techniques can be introduced to undertake mechanical operations and quality assurance testing.

Through careful preplanning and the adoption of standardised designs, the costly effects of post-construction design changes should be eliminated. Given these practices and the establishment of good labour relations, it has been shown to be possible to build 1 000 MWe nuclear plants in as little as five years (less in some cases). French 900 MWe plant experience is illustrated in Figure 16. This contrasts with less fortunate examples, in which countries have experienced reactor construction periods of ten years or more. This adds greatly to the magnitude of interest charges during construction and the overall capital costs.

f)   Reduced commissioning period

The rate at which new plants can be commissioned and brought up to full power is important because earnings from electricity sales in the early years of a plant's life contribute more in present-worth income terms than equivalent earnings later in its life (see Chapter 5 and note 3 of Annex).

Some reactors have taken several years to achieve their full design output. However, experience with water-cooled reactors has been good, and the time taken to ascend to full

Figure 17. **Time to full power - German experience**

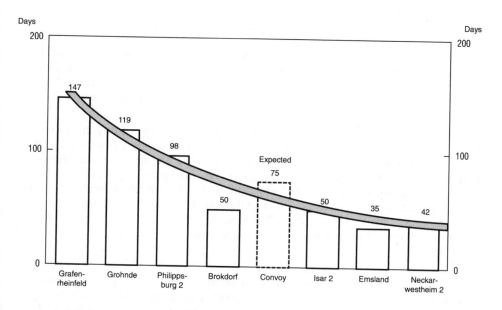

Source: *Good Performance in Nuclear Projects,* NEA, 1989

power has been reduced dramatically for proven designs. Figure 17 is an example of German experience.

Reactor design

a) Advanced water reactors

Since water-cooled reactors, in one form or another, now constitute over 90 per cent of all nuclear plants in the WOCA area, and this dominance seems likely to persist for some time, it is not surprising that designers continue to look mainly at this type of plant for technical improvements and improved cost-effectiveness. Designs incorporating such improvements are often given the generic title of "advanced reactors". The term however covers a range of developments, running from relatively modest evolutionary changes from existing reactor designs to more radical rethinking of the whole plant structure and its mode of operation.

Exchanges of views and information on the rationale for and current status of advanced water reactor designs took place under the aegis of the Nuclear Energy Agency in 1988/89. In practice, all countries with nuclear programmes were actively seeking to apply more advanced operating technologies to existing reactors. These included use of extended burn-up fuels to improve reactor and fuel utilisation and, in some countries, the recycling of plutonium in mixed-oxide fuels. Improvements continue to be sought in plant reliability by making use of advanced control systems and greater consideration of human factors. Advanced maintenance and loading technologies have been developed to reduce the occupational exposures of workers to radiation and to minimise the volumes of low-level radioactive waste.

Advanced reactor designs have been adopted by many countries for their current construction programmes. These are evolutionary developments of the earlier plants that incorporate improvements in the man-machine interface and make use of the latest materials technology for pressure vessels, pressure tubes and heat exchangers. Designs have been improved to simplify construction, operation and maintenance. These developments are expected to yield significant benefits in terms of greater plant availability and lower capital costs.

The advanced plants that are commercially available are mostly large-capacity plants in line with the size trends that were witnessed in the 1980s. Canadian pressurised heavy-water reactor designs of 300 MWe and 600 MWe are the exceptions. However, smaller advanced light-water reactor designs are also under development in the United States, Sweden, Japan and the United Kingdom. Some of these rely basically on current technology but with increased use of modularisation and greater reliance on passive safety features. Others are more radically innovative and incorporate novel engineering concepts that may require operational testing in large prototypes prior to commercial licensing.

The overall objective of reducing the costs of electricity production from nuclear plants, while maintaining or improving their safety, is shared by all reactor designers. First priority has always been given to assuring the safety of the plant and minimising all possible environmental impacts. Subject to this requirement, the reduction of basic construction costs through design improvements and simplification, and through the introduction of design features calculated to improve overall system reliability, are the overriding goals. In general, design changes aimed at simplification and improved reliability will also improve the overall safety characteristics of any reactor.

It is an essential prerequisite of design improvement that there should be effective communication and feedback of experience from reactor operators to the designers and, wherever possible, that maximum use be made of accumulated international experience in the fields of conceptual design, choice of materials and instrumentation and control procedures.

There is confidence that the advanced large-scale water-cooled reactors now being brought forward will have benefited from accumulated international experience, and that they will offer the prospect not only of significantly higher reliability (and availability), but that their specific capital costs (cost per kW) will be significantly lower than those of earlier designs of similar power rating. Thus in Japan, an advanced boiling-water reactor is expected to cost some 20 per cent less than its predecessors.

b) Small and medium power reactors

As noted previously, the trend in most OECD countries has been towards the design and construction of progressively larger thermal reactors, with the most recent designs approaching 1 500 MWe capacity. The driving force behind this trend has been the expectation that the specific costs of plants (i.e. cost per kW of capacity) would be reduced as a result of the well-established scaling laws related to engineering projects. However, there are also disadvantages that arise from the move towards larger plants. The market for such plants is limited to those countries having access to an interconnected grid with total power requirements at least 10 to 20 times the capacity of the largest plants used to supply it, otherwise the reliability of the grid can be affected. This rules out their deployment in a large number of countries outside the OECD, unless they are connected to an international interlinked system with a high combined capacity, such as those serving the regions of Europe.

The adoption of very large plants also necessarily involves the consequence that problems with any one plant can result in major losses of output, relative to those that might be expected if the same capacity were to be provided by a number of smaller plants of similar individual reliability. This leads to a need for larger margins of spare capacity and increases the cost to the utility.

Small reactors offer the opportunity of better matching increments of generation capacity to the growth of electricity demand and could, in some circumstances, provide an easier route to replacing existing smaller fossil-fuelled or nuclear plants when they have reached the end of their operational life. Finally, it is argued that smaller reactors offer greater opportunities for modularisation and factory construction of components, with the consequent minimisation of costly, weather-dependent, site work.

In recent years, there has been the additional thought that it would be possible to design smaller reactors which placed great emphasis on passive safety systems. Such systems might be more readily understood by the general public and could therefore facilitate public acceptance of nuclear plants in those countries where concerns over reactor safety, no matter how exaggerated, were acting as barriers to the deployment of nuclear power.

The stress currently laid on the need for high standards of safety and the greatest possible transparency of the ways in which such safety is guaranteed, has led several countries to look for simple rugged designs with increased safety margins. This would permit longer periods of time before human intervention is required in the event of any plant failure, it would reduce risks of damage to the reactor core and might reduce further the already very small risks of accidents affecting the general population.

As yet there are no small and medium reactors of modern design in operation, although some 15 designs for use in electricity production have been produced in OECD countries, along with seven designs directed specifically at heat applications. All of these designs share the common objective of simplification of the safety systems, through the application of such measures as the adoption of passive emergency cooling and passive residual heat removal systems. Such measures can permit the elimination of the more complex safety systems associated with the large-scale water-cooled reactors. The resulting cost savings help to offset the loss of benefits of scale that arises from the move to smaller plants.

Additionally, the nuclear heating plant designs share the characteristic that they have very low core power densities. These are associated with undemanding temperature and pressure requirements and permit further simplification. In principle, nuclear heating plants can be designed to operate, should it be desired, without a need for any resident operators, and sited in close proximity to centres of population.

Analysis of past experience of reactor operation suggests (Table 10) that the availability factors for small plants could well exceed those of their larger counterparts, due to their greater simplicity and their smaller number of components. Studies in a number of OECD countries suggest that the small and medium reactors could produce electricity at costs that are not widely different from those of large plants of 1 000 MWe capacity or

Table 10. **Effect of scale on US reactor performance**

Performance of Westinghouse PWRs

Size range (MWe)	No. of units produced	Cumulative operating experience (reactor years)	Cumulative average load factor (percentage)	Annual load factor for 1987 (percentage)
<600	16	270	68.8	77.8
600-999	32	257	61.2	67.1
>999	25	177	58.2	60.8

Performance of General Electric BWRs

Size range (MWe)	No. of units produced	Cumulative operating experience (reactor years)	Cumulative average load factor (percentage)	Annual load factor for 1987 (percentage)
<600	9	151	60.2	67.2
600-999	23	278	58.3	63.6
>999	17	119	51.8	46.4

Source: Small and Medium Reactors, NEA, 1991.

higher, when the benefits of design simplification, of modularisation and replication, and of reduced construction time are taken into account.

Figure 18 illustrates one Japanese view on the potential effects of the different factors on overall investment costs, for a series of small reactors of 600 MWe substituting for 1 200 MWe reactors at three different rates of demand growth and interest rates. The factors include replication of units, simplification of design, construction period reductions, and improved capacity factors (availability). Offsetting costs arise from higher operation and maintenance costs because of the larger number of units and the losses of scale benefits (not represented).

Table 11 presents analogous data in the form of an overall generation cost comparison based on British paper studies. They suggest that smaller reactors (SIRs, or Safe Integral Reactors) might be able in due course to produce electricity at costs comparable to those of large presssurised water reactors.

There remains a problem however, in that the advanced small reactor designs, particularly those based on more novel concepts, have yet to be demonstrated. Prototypes need to be built and operated before their economics and licensability can be established. At the present time, the market for new nuclear plants is very depressed, and those countries that are proceeding with significant programmes of investment are committed to the large designs that they have already shown to be satisfactory, and with which they have considerable operating experience.

Those countries with small grids capable of accepting only small plants do not, in general, have the infrastructure to undertake their own reactor development. Nor do they have the capital that would enable them to underwrite development of small plants, for what might still turn out to be very limited applications.

It seems most likely therefore that the best prospects for small and medium reactor development rest with those advanced industrial nations whose nuclear programmes have faltered because of public opposition. In such countries, the greater transparency of the safety features of some smaller reactor designs may assist in overcoming public concerns and facilitate the reacceptance of a significant nuclear contribution to the countries' future electricity generation infrastructure.

The multiplicity of design concepts for small and medium power reactors is itself a barrier to their early acceptance, and has parallels with the range of radically different designs that were pursued in the early days of nuclear power development. The problem arises from the fact that the existence of a range of designs, which might be offered on the world's markets, reduces the prospects of recovering the considerable development expenditure required for each individual plant. Potential purchasers are also likely to hold back to see which designs gain widest acceptance before committing themselves. For these reasons, amongst others, there have been moves to internationalise and consolidate the reactor designs, focusing on a smaller number of concepts.

Should small and medium reactors be fully developed, they have prospects both of improving public acceptance in those countries where problems currently exist, and of increasing the number of countries able to deploy nuclear power effectively within their national networks. This would increase the overall contribution nuclear power can make to world electricity supplies. Small reactors also offer the prospect of enabling nuclear power to enter the heat supply market, through use of small water-cooled plants providing low-temperature heat for industrial, commercial and domestic purposes. Additionally, modular high-temperature reactors could contribute to high-temperature industrial

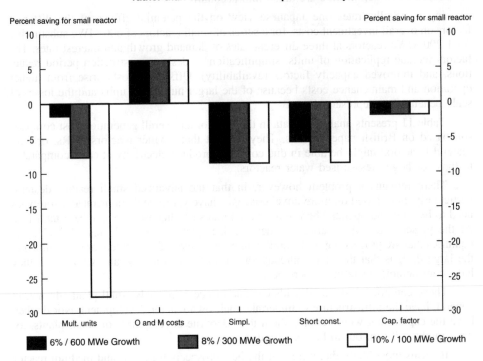

Figure 18. **Savings arising from construction of 600 MW units rather than 1200 MW units - Japanese study**

Percent saving for small reactor

| | Mult. units | O and M costs | Simpl. | Short const. | Cap. factor |

■ 6% / 600 MWe Growth   ■ 8% / 300 MWe Growth   □ 10% / 100 MWe Growth

Note: At different interest rates and annual capacity growth

Source: Small and Medium Reactors, NEA, 1991

Table 11. **United Kingdom relative specific costs for PWRs of different capacity**

	1 200 MWe	600 MWe	300 MWe	150 MWe
Capital cost	0.66	0.82-0.94	1.04-1.38	1.32-2.07
Savings for small plant	–	0.1-0.2	0.27-0.41	0.41-0.63
Adjusted capital cost	0.66	0.62-0.84	0.63-1.11	0.74-1.66
Fuel costs	0.21	0.21	0.21	0.21
O and M	0.11	0.14-0.17	0.19-0.25	0.25-0.39
Decommissioning	0.02	0.02-0.03	0.02-0.05	0.02-0.05
Total generating costs	1.0	1.0-1.3	1.1-1.6	1.2-2.3
Other savings[a]	–	0.1	0.2	0.2-0.3
Potential overall cost	1.0	0.9-1.2	0.9-1.4	1.0-2.0

a) Better demand matching, planning margin savings, learning, etc.
Source: Small and Medium Reactors, NEA, 1991.

processes such as steel making, hydrogen production or substitute natural gas manufacture.

However, it still remains to be demonstrated that under current world economic circumstances, small reactors will prove economically competitive either with their larger counterparts, or with the production of electricity and supply of heat from small fossil-fuelled sources and, when available, renewable sources. Nevertheless, current concerns about the environmental implications of the growing use of fossil fuels and the inevitable growth in the energy requirements of developing countries, could materially improve the prospect for smaller nuclear plants.

c) *High-temperature reactors*

The predominant water-cooled reactors are restricted to coolant temperatures of around 300°C by the properties of the coolant. This limits their thermal efficiency in power generation to levels well below those achieved in modern fossil-fuelled plants. This is not a major problem for power plants, but the achievement of higher temperatures would help to increase electricity output for a given investment, and would offer the prospect mentioned earlier of using nuclear heat in high-temperature manufacturing processes.

British advanced gas-cooled graphite-moderated reactors (AGRs) were a step in this direction with carbon dioxide coolant temperatures of 350°C to 650°C, comparable to those achieved in liquid-sodium-cooled fast reactors. However, high-temperature gas-cooled reactors (HTRs) using inert helium gas as coolant have been designed to reach temperatures as high as 750°C to 950°C.

The demonstration reactors built to date have been of small to medium power, and it is in this context that they have been considered in Nuclear Energy Agency studies. They are well-suited to modular construction and offer many of the advantages described in the previous section. They also have high negative coefficients of reactivity: i.e. if their power rises, the core temperature increases and causes changes in the material character-istics, which then reduce the power level, thus providing a degree of self-stabilisation. The large mass of graphite moderator also helps to slow down the rate of temperature rise compared with water-cooled systems, in the event that coolant pumps fail.

The fuel is a stable high-temperature ceramic with no metallic cladding to corrode, and the use of mixed uranium-thorium fuels can give high fuel burn-up by converting some of the thorium into fissile uranium-233 *in situ.*

The pace of HTR development has slowed during the past decade since the potential markets have not materialised, and the need for high-temperature processes, such as hydrogen production and substitute natural gas manufacture, has not existed at a time when fossil fuels have been plentiful and cheap. Nevertheless the technology of the HTR has been amply demonstrated and it remains a prospect for the future.

d) *Advanced designs for improved fuel utilisation*

Several reactor designs are under development within OECD countries with the specific aim of improving fuel utilisation rather than the previously described aims of reducing the capital costs or increasing safety margins.

Some of these reactors are water-cooled reactors that aim specifically at getting higher burn-up of fuels and thereby reducing overall fuel cycle costs. Amongst these can be mentioned the French variable neutron spectrum reactor, the Germano-Swiss high-conversion LWR, the heavy-water-moderated advanced boiling-water reactor (the ATR) being developed in Japan, and the Canadian concept of using light-water and heavy-water reactors in tandem. In the latter system, spent LWR fuel, with too low a uranium-235 concentration to stay in service in LWRs, can be used as fuel for the more neutron-efficient pressurised heavy-water reactors that currently operate with natural uranium oxide fuel.

With the present abundance of uranium on world markets, there is little incentive to press forward with developments aimed at uranium conservation, and the attractiveness of the different reactor concepts will depend significantly on the extent to which they can cut the overall cost of fuel fabrication and fuel management.

Meanwhile, continuing attention is being given worldwide to the development of plutonium-fuelled fast-neutron reactors (FRs) cooled using liquid sodium (Chapter 3). These reactors are under development in France, Japan, the United Kingdom, United States and the former USSR.

The principal attraction of the fast reactor is that it can take depleted uranium, either from enrichment plant tails or the material recovered by reprocessing spent thermal reactor fuel, and convert it into plutonium using the excess neutrons produced by plutonium fission in a fast-neutron flux. This excess arises from the fact that in the absence of the neutron-absorbing moderator (water), present in systems like the light-water reactor, fewer neutrons are required to sustain a fission chain reaction in the plutonium fuel.

In principle, the fast reactor can produce (breed) enough fresh plutonium to sustain its own operation indefinitely, together with a small additional quantity that can be recovered and used to provide the initial fuel for additional fast reactors. By means of this conversion process, the quantity of energy recovered from natural uranium can be increased by some 60-fold. This multiplication of the energy recoverable from uranium has to be achieved if nuclear power is to make more than a minor and transitory contribution to the world's energy supplies (see Chapter 4).

Fast reactor fuels have been demonstrated to be capable of achieving far higher burn-ups than have yet been attained in thermal reactors. This greatly reduces the quantities of fuel that have to be fabricated and subsequently reprocessed in order to recover the plutonium needed to sustain the fast reactor fuel cycle. This, together with the removal of any need for fresh uranium and enrichment services, more than offsets the extra costs involved in handling plutonium fuels, and can result in significant fuel cost savings for fast reactors, relative to their thermal reactor counterparts.

Unfortunately fast reactors, as currently designed, rely on a double set of heat exchangers. They use the radioactive sodium coolant from the core to heat unirradiated sodium, which in turn transfers the heat to water in steam-producing heat exchangers. This design complication adds to the capital costs of fast reactors and, so far, their capital costs have been significantly higher than those of large-scale light-water reactors.

Current design effort is directed towards reducing the capital cost differential to a point where uranium-fuelled thermal reactors and plutonium-fuelled fast reactors have comparable electricity generation costs. Programmes in Europe, Japan and the former USSR are all developing fast reactor designs based on the use of plutonium-uranium oxide fuels backed by conventional fuel reprocessing technology, in which the fuel is dissolved and chemically processed in solution to recover the unburnt uranium and plutonium for recycling.

The United States has developed a design for a modular fast reactor using metallic plutonium fuels in which high-temperature techniques, without the use of chemical solvents, are used to remove fission product impurities and recover the uranium and plutonium for recycling in fresh fuel.

The oxide-fuelled fast reactor technology and its fuel cycle have been developed and demonstrated at the commercial scale. It is, however, now generally accepted that significant commercial deployment of the technology will not need to take place for some time, unless there is a massive expansion in the capacity of thermal nuclear plants in operation worldwide. This is a consequence of the slower than expected growth of nuclear capacity, and the projected availability of uranium resources to fuel thermal reactors well into the next century. This leaves time to improve the design to reduce capital costs and to overcome the technical teething problems that have been encountered with the non-nuclear sections of prototype and demonstration fast reactor plants.

This does not mean, however, that the development of fast reactors can be put on one side until they are needed. The availability of proven commercial designs in the future depends critically on their continued evolutionary development and demonstration, for which little enough time exists should there be an upturn in nuclear electricity demand.

Fuel cycle improvements

The uranium oxide fuel elements used in the majority of thermal reactors have been designed to ensure efficient heat transfer from the ceramic fuel through the cladding to the reactor coolant; to accommodate dimensional changes in the fuel during irradiation; and to contain the fission products within the sealed fuel cladding. The fuel is manufactured to very tight tolerances with rigorous quality control, which has helped to ensure that the number of fuel pins developing leaks within the reactor core (and releasing gaseous fission products into the core) are maintained at negligibly small levels (Table 12).

Systematic development has improved all aspects of fuel performance including its burn-up, that is the amount of energy recovered from the fuel before it has to be removed from the core. Typically, today's light-water reactors achieve burn-ups of some 33 GWe-days per tonne, but advanced thermal reactor fuels have been developed and demonstrated that are capable of giving burn-ups in excess of 50 GWe-days per tonne. The object of these improvements is to reduce the overall costs of the fuel cycle and electricity production from nuclear sources.

Higher burn-up fuels require higher enrichment with uranium-235 and a larger quantity of natural uranium feedstock. Against this, their fabrication costs and reprocessing costs per kilogram of fuel remain unchanged, and the net fuel costs per unit of energy produced from the fuel decrease. Figure 10 illustrates this effect.

An incidental benefit arising from the availability of higher burn-up fuels is the ability to refuel at longer time intervals, or to refuel at the same interval but to replace a smaller portion of the core fuel at each reloading. Either of these alternatives can increase

Table 12. **United Kingdom fuel performance**

Year	Number of pins irradiated, to nearest 1 000 (cumulative)	Number of pin failures	Failure rate
1976	320	0	
1977	330	0	
1978	365	1	2.7×10^{-6}
1979	415	1	4.8×10^{-6}
1980	495	0	
1981	555	0	
1982	625	0	
1983	955	0	
1984	1 205	1	2.5×10^{-6}
1985	1 385	0	
1986	1 470	0	
1987	1 610	0	
1988	1 930	0	1.5×10^{-6}

Source: Good Performance in Nuclear Projects, NEA, 1989.

the overall availability of the reactor and enable it to reduce its unit costs by increasing the output for the same fixed capital investment.

In addition to the development of improved uranium oxide fuels for thermal reactor use, development has proceeded on mixed plutonium-uranium oxide fuels (MOX) for use in existing water-cooled reactor plants (see Chapter 5). Plutonium is an inevitable by-product of the use of uranium fuels in thermal reactors, but its use in fuels has raised fears in the minds of many because it is highly toxic when inhaled, ingested or when it enters the bloodstream through a wound. However, 30 years of experience have shown that it can be safely managed and used as a nuclear fuel, subject to appropriate precautions, and that it poses no greater threat to the public than many other materials utilised industrially.

It has also been demonstrated that MOX fuels, designed to have the same fission characteristics as uranium oxide fuels, can be used as a substitute in the refuelling of many light-water reactors up to proportions of around 30 per cent, with no adverse consequences for the safety of reactor operations. MOX fuels are already coming into more widespread use in thermal reactors (Chapter 4), and development of higher burn-up MOX fuels is being undertaken in parallel with that of uranium oxide fuels.

Technical development has continued in all the separate segments of the nuclear fuel cycle. Particular attention has focused over the years on the uranium enrichment process, which is a major element in the cost of uranium fuel. The energy-intensive diffusion process, widely adopted in the early days of civil nuclear power, continues to be the predominant process in the United States and France. The United Kingdom, the Netherlands and Germany collaborated to develop the high-speed centrifuge process, which is much less energy intensive. Recent attention in several countries has focused both on improvements to this latter technology and on the development of laser-based separation techniques, which have been demonstrated to be practicable but which have not yet reached the stage of commercial deployment.

Steady if unspectacular progress is being made in other areas of the fuel cycle, including reprocessing technologies, spent fuel and waste storage, waste conditioning and encapsulation.

Overall performance

Every sector of the nuclear industry is striving to improve all aspects of its performance to match the best international experience. Some countries and companies are further along the path to all-round technical excellence than others, but even those with the most successful programmes continue to look for ways of reducing costs and building on their success.

There is some confidence that the basic costs of reactors and their fuel stabilised during the 1980s, and that the benefits of technological improvements will begin to be increasingly apparent in the form of real generation cost reductions during the 1990s. This would provide a good base for the wider exploitation of the technology in the post-2000 period when a major programme of generation plant replacement will be beginning.

Chapter 7

TECHNOLOGICAL AND ECONOMIC CHALLENGES AND OPPORTUNITIES

Present situation

Nuclear power, which was one of the fastest-growing industries in the world in the 1970s, has become a major contributor to the world's energy supplies as described in Chapter 2. At the same time, its technology has not stabilised but has continued to evolve, although attention has now concentrated on a limited range of technologies that are believed to have the greatest commercial potential in the short and medium term.

Nuclear technology appeared during the 1970s to be gaining a considerable economic advantage over the competing fossil fuel options, except in places with direct access to low-cost coal supplies, but it has failed to capitalise on this promise in many countries. This has been due in part to the increased costs that nuclear plants have had to bear in order to provide assurance over their safety, and partly due to the fact that the world economy, world energy demand and fossil fuel prices have not grown in the way that was at one time anticipated. Nevertheless, nuclear power has remained competitive with fossil fuels for baseload power generation throughout most of the OECD and has had a significant edge over its competitors in those countries where there have been consistent programmes based on replicated designs.

Despite the success that nuclear power has had, a significant proportion of the population in some OECD countries is strongly opposed to it deployment, and its use has become a major political issue in a number of states. There are some signs that the new-found political and public appreciation of the potential consequences of continued world reliance on fossil fuel combustion for the bulk of its energy supply, has mellowed attitudes towards nuclear power and brought about a greater appreciation of its broader beneficial effects.

In order to build up public confidence and fully justify it being allotted the major role it has the potential ability to fulfil, the nuclear industry has to provide clear and easily understandable evidence of its safety. It has to demonstrate that its potential impacts on the environment are trivial and that it has considerable advantages over other conventional sources of energy in this regard. Finally it has to satisfy the world's electricity utilities that its costs are well-defined and that it will remain competitive economically with other energy technologies in the future.

From the standpoint of the Nuclear Energy Agency's Nuclear Development Committee, there are a number of clearly definable areas where international co-operative studies can help. They can do this by establishing expert consensus on the current position and setting out clearly the remaining uncertainties that could affect the technical and economic position of nuclear power in the future. Some of these are discussed below.

Nuclear capital costs

Since capital costs are the biggest single contributor to nuclear electricity generation costs, it is imperative that the nuclear industry makes every effort to contain or reduce them, subject always to maintaining the very highest standards of safety.

In Chapter 6, the different routes available to achieve this goal have been described. For some OECD countries, the mere establishment of a coherent programme of reactor construction based on internationally accepted, readily licensable and replicated designs would greatly improve the economics of future nuclear plants relative to those they have so far constructed, particularly if collocation of a number of reactors at a single site can be accommodated.

Other countries have already derived considerable cost savings by means of these measures and, for further reductions, will have to look to the simplification of design and the steady improvement of performance, both in the construction and the operation of plants. In these areas, continuing exchange of international experience, the careful monitoring of performance, good staff training and detailed advanced planning using the most up-to-date management techniques all have a role to play.

Plant life extension

Because the capital cost of nuclear plants is high and the fuel and operating costs are relatively low, it is important that the maximum possible use is made of any existing nuclear plant. As described above, the prime objective of plant operators is to get the very best availability from their existing plants by minimising plant outage times (both planned and unplanned).

However, a new opportunity is arising for the efficient use of capital resources. Many nuclear plants are coming towards the end of the lives originally planned for them. These lives were based on conservative technical assumptions and the adoption of considerable margins of safety, based on the state of knowledge and experience that existed 20 or more years ago. It is now realised that many of the existing reactors could continue to operate safely for periods significantly in excess of their original design lives. If this is indeed the case, then the extra output of electricity becomes extremely cheap relative to that from other generation options.

Clearly the national regulatory authorities have to be satisfied that extended operation is safe. They will be expected to examine very carefully whether any plant refurbishment, component replacement, etc. is necessary in order to provide assurance of continued safe operation. Additionally, the operating regime of the reactor, e.g. its power rating, may need to be modified.

Increasing international consideration is being given to the question of what might be involved in extending the life of existing plants, both in terms of any necessary additional investment and its timing, and whether the management of the reactor and its fuel cycle during its normal scheduled life can affect the feasibility or the costs of life extension. There appears to be no inherent reason why water reactors can not have their lives extended to well beyond the 40 years that some designers now project. Even this contrasts with the 20 to 25 years many countries have conservatively assumed in their earlier planning.

Decommissioning

International technical exchanges have established expert consensus that reactor and fuel plant decommissioning poses no insurmountable problems, and have demonstrated that the techniques already developed for maintenance and smaller-scale plant decommissioning are capable of dealing with large commercial reactors and facilities.

There is still some public scepticism, however, and some divergences of view exist within the industry about the overall costs of decommissioning large nuclear plants. The reasons for this are now well understood. Some differences arise from different national policies regarding waste management, or from different expectations of what future policies might be. Despite these differences, it is unambiguously clear that current expectations of decommissioning costs would have to be grossly in error before they could significantly affect the overall costs of electricity production using large water-cooled reactors.

Continuing exchanges on the evolving experience of the decommissioning of existing facilities, which is already in progress in a number of OECD countries, will assist in generally improving the techniques and establishing common standards. They will also enable the critical comparison of different national approaches and requirements, and will provide assurance that no unanticipated problems have arisen that could alter perceptions of the significance of decommissioning as a factor in overall nuclear economics.

The nuclear fuel cycle

Developments of nuclear fuel technology are largely taking place within private sector or publicly owned corporations that are in commercial competition with each other. Under these circumstances, the role that international agencies can play is limited, except in relation to safety and regulation. Nevertheless, it has proved valuable to gather data on the levels of uranium resources and the existing and planned capacities of fuel cycle plants. This data has been and continues to be useful to governments and private industry, through the provision of a reliable basis for forward planning.

It has also proved constructively useful to develop international consensus on the general future costs of nuclear fuel services and nuclear fuel, both to underpin studies of overall generation costs and to provide a reliable expert consensus on the likely future costs of nuclear fuels.

Such studies serve to highlight areas of uncertainty in the nuclear fuel cycle and to identify those components of cost to which the overall fuel cost is most sensitive. Because market circumstances change, because technologies evolve and because the regulatory framework can be revised, with consequences for the costs of some fuel cycle processes, there is need to update periodically any studies dealing with these costs.

Apart from uranium prices, which have been radically affected by the depressed market circumstances of the 1980s and early 1990s, there have been significant uncertainties in some components of the back end of the nuclear fuel cycle in particular. These are considered below.

Spent fuel stores and nuclear waste repositories

Unlike most stages of the nuclear fuel cycle, there has, as yet, been no commercial market for the storage of spent nuclear fuel or for the disposal of that fuel or of reprocessing wastes. For this reason, the costs of these stages of the back end of the nuclear fuel cycle have had to be estimated on the basis of paper studies conducted in individual OECD countries.

A wide range of different techniques is available for the storage of the spent fuel and its encapsulation prior to disposal. There is also a wide range of designs for geological repositories using clay, salt or volcanic rock matrices. It is inevitable therefore that there will be a significant range of possible future costs. Once again this should not be interpreted as revealing ignorance of costs on the part of the industry. For a given and clearly defined requirement, concerning the fixation and immobilisation of wastes, appropriate facilities can be designed and costed with some confidence. Even with the uncertainty about the precise specifications that will be required, the existing uncertainty revealed by international experience is not such that it has a significant influence on overall nuclear generation costs.

The development of international expert consensus setting out clearly the range of options and the spectrum of costs that might arise, given various regulatory requirements, can help to allay public misconceptions about the extent of industry knowledge in this area.

New technologies

The majority of new technologies are being developed in a commercially competitive environment, and this has complicated the exchange of information outside the framework of specific co-operative or joint project arrangements.

Enthusiasm within the nuclear industry for the development of radically new technologies has been considerably dampened as the costs of development, licensing and demonstration have risen, and the potential markets perceived for new plants have diminished. This has led to a focusing of attention on the improvement of existing plants and modest evolutionary changes based on prior experience.

The principal exceptions to this have been small and medium power reactors, where major new market opportunities have been considered possible, and the continuing development of fast reactors, whose deployment is seen as inevitable and essential if nuclear power is to make a long-term major contribution to world energy supplies. International exchanges on relevant nuclear and technical data continue with mutual benefit. Useful exchanges have also taken place on general design concepts, and views on appropriate specifications and market opportunities can be and have been assembled on a generalised basis.

A major concern within the nuclear industry has been the establishment of means of financing technologies with long development times before commercial benefits can be realised. Costs of many advanced technological projects, not only those in the nuclear area, are high because they have to be developed and demonstrated at full scale to give the necessary degree of assurance concerning their safety, reliability and economic prospects. In the past this problem has been less acute because governments either singly or via co-operative programmes have backed major developments. The tendency in recent years for governments to look more towards private industry to support advanced development raises major problems for high-cost long-running development projects with delayed benefits.

One role for an international agency in such a situation might be to seek to catalyse industrial co-operative arrangements so that development costs falling on individual companies are reduced and the potential markets arising from development shared. Another useful service can be the objective assessment of the need for and benefits that might potentially arise from the development of the technology (if any), and analysis of the means by which any such benefits could be derived. Such analysis would assist OECD countries, either singly or in concert, in making decisions on the need for, and preferred routes to the achievement of specific technological development goals.

Scientific exchange

The nuclear industry has been noted for its willingness to exchange basic data on nuclear properties, on safety-related matters and on scientific matters. On many subjects, exchanges take place under the aegis of the International Atomic Energy Agency in Vienna. Others take place under the OECD Nuclear Energy Agency.

A current example of the latter is the information exchange programme on actinides. These are the transuranic elements produced by successive neutron capture in uranium and the other heavy elements produced *in situ* in nuclear reactor fuel. Although a great deal of information exists on many of these elements, on their isotopes and their decay products, it has been recognised that there can be mutual benefit in further exchanges concerning their nuclear properties and characteristics. These are significant in relation to their behaviour in high burn-up plutonium fuels and the consequences of their presence in nuclear wastes.

Exchanges also focus on techniques for separating the actinides from fission product wastes and their transformation to short-lived products using particle accelerators or fast reactors. A considerable body of knowledge exists, but the development of suitable technology is still seen as being a lengthy process.

Broader economic impacts

Most decisions on the preferred options for future energy investments have been based on the straightforward financial comparison of these options, taking into account the possible future evolution of fuel prices and any possible concerns about the security of fuel supply. Such concerns, arising from whatever cause, have been considered by some governments and utilities to justify the payment of a small premium to ensure diversity of plant types and fuels used within a national electricity supply network. The wisdom of having a diversified fuel base has been highlighted by the crises that have occurred in the Middle East, with their impacts both on availability and price of hydrocarbon fuels, and by the effects of strikes in the fuel industries on the ability of utilities to maintain their electricity output.

These and many other factors have been cited from time to time as grounds for establishing or maintaining a diverse base for the electricity supply industry. The existence of broader benefits has also been seen by some as an additional argument that could help to offset public concerns about the deployment of nuclear plants. Postulated benefits have included improvements in a country's balance of payments, stabilisation of fossil fuel prices, contributions to national economic growth (over and above those that would be achieved from investment in fossil fuel technology), and the avoidance of deleterious environmental consequences associated with fossil fuel combustion.

With current international attention being focused on the consequences of the greenhouse effect and on determining the need to reduce greenhouse gas emissions, there is a renewed international interest in establishing a consensus on the benefits and costs, over and above those reflected in conventional investment appraisal, that nuclear power gives to or imposes on society at large. The Nuclear Energy Agency is currently co-operating with some of its Member countries and with other international organisations in studies directed towards establishing, in so far as possible, what is known about the reality, magnitude and significance of these costs and benefits, and of the methods that have been deployed in OECD countries to seek to quantify them.

An NEA study to be published in 1992 shows that conventional microeconomic cost analyses, such as those discussed in Chapter 5, do reflect the full costs of nuclear electricity supply, and that the external environmental and health costs imposed on third parties by nuclear power, both in routine and hypothetical accident conditions, are very small in relation to the direct generation costs. The same claim cannot be made for fossil-fuelled plants, even when they are fitted with desulphurisation and denitrification technologies to limit acid gas emissions, since the social costs of carbon dioxide emissions could be several per cent of generation costs or considerably higher. There is still great technical and economic uncertainty about the implications of greenhouse gas releases to the atmosphere. Where nuclear power is the cheapest baseload generation option, its use can contribute to a small increase in gross domestic product, to higher employment and to an improved balance of payments. On a global scale, its use helps to stabilise fossil fuel prices and its abandonment would have a major worldwide economic impact, comparable in scale to the oil crises of the 1970s.

Parallel studies are also in hand to bring together information in as quantitative form as possible on the so-called spin-off benefits that have arisen from the work of the nuclear industry. These are the benefits not directly associated with the deployment of nuclear power, but which accrue to other sectors of industry and commerce as a result of the nuclear industry's activities.

The identification of historic spin-off benefits has no necessary relevance to the likely benefits that may flow from further nuclear development. However, there is general recognition of the fact that it is important to ensure that the knowledge, skills, materials and techniques developed in the course of nuclear or other advanced technology programmes, should be capitalised on to ensure that the maximum benefit for national economies is realised. The transfer of technology from one sector of the economy to another is not easy however, and exchange of experience concerning the methods that are used and their relative success can be beneficial to all participants.

Constraints to nuclear development

Apart from the obvious question of public attitudes and their influence on the acceptance and adoption of nuclear power, there are other potential constraints that could influence the ability of individual countries to undertake nuclear programmes that they consider to be desirable.

These constraints could arise from the limited availability of qualified and trained manpower to undertake the design, specific construction or operational tasks. There is also a possible constraint that could be imposed by the inability of the design and construction industries to undertake tasks on the scale or at the pace felt to be necessary.

If the nuclear industry had reached a stable state in all OECD countries with a regular pattern of plant ordering and construction (or one which was growing slowly but steadily), there would be no problem, because the institutional mechanisms would have been established that would ensure a continuing supply of new entrants to the industry to maintain its strength and capabilities. However, this is not the situation that has existed or is likely to exist in a number of OECD countries. Moratoria on future plant construction, either as acts of policy or as a consequence of economic circumstances, have led to significantly reduced rates of plant ordering and construction, to zero in the limit. This has resulted in a number of design and construction companies withdrawing from the nuclear field over the years, and this clearly has an adverse effect on recruitment of high-calibre staff to the industry. Whether or not this is significant depends on whether countries believe that they will need to reinstitute or accelerate their programmes at some future date.

The Nuclear Energy Agency is in the course of examining the situation in OECD countries, more especially in relation to the availability of trained manpower. Information on this topic can help to underpin future policies for education and training and to identify the need, if any, for steps to strengthen the relevant industrial infrastructure.

Chapter 8

THE ROLE OF AN INTERNATIONAL AGENCY

This brief review has set out the work that has been undertaken under the auspices of the Nuclear Energy Agency's Committee for Technical and Economic Studies on Nuclear Energy Development and the Fuel Cycle in recent years. Some of this work is ongoing and will appear in the form of published reports in coming months. A list of publications and studies is provided in the bibliography at the end of the text.

There are perhaps three main functions that the Agency can usefully fulfil in the area of the technical and economic development of nuclear power. The first is to act as a forum where experts drawn from OECD countries can exchange technical data and experience and, where appropriate, co-ordinate their research, design, development and other activities to minimise the costs falling on individual countries. The second function is the development of international expert consensus on the current status of technologies or the current economic position of nuclear power, its fuel cycle or related technologies. The dissemination of this consensus to governments and a wider public audience can assist in creating a knowledgeable dialogue on policy issues. The third function is the investigation of other topics related to nuclear power that have a particular policy relevance, the assembling of related data, and the review of their implications. Examples of the latter include ongoing studies of the availability of trained manpower and the methods of accounting for and financing liabilities arising from current and past nuclear activities, such as nuclear wastes and radioactive plant decommissioning.

It is clearly important that studies are conducted in an objective manner and that the negative as well as the positive aspects of nuclear power are fully brought out. To do otherwise diminishes the value of studies and could seriously reduce their impact.

It is not sufficient, however, to undertake studies and produce reports. It is essential that the results of studies are made available to all who could benefit from using them in a form that is easily understandable, and which brings out the relevance of any findings to the types of decisions and choices that are being faced in OECD countries. In this regard, it is not enough for the nuclear industry itself to be convinced that it has the technologies and reasonable costings, for example for plant decommissioning and waste disposal. This message, and the arguments supporting it, have to reach the wider public and the media if broadly based understanding is to be achieved and rational debate conducted.

This is particularly important at the present time when energy policies and nuclear power are both at a crossroads. Energy policy, because of the dilemma posed by the growing needs of the third world and the question mark over the sustainability of even current energy consumption, in the light of environmental and ecological concerns.

Nuclear power, because of the de facto moratoria on its use that are in place in many OECD countries, at a time when it appears to its proponents to offer an environmentally attractive large-scale supply option that is capable of overcoming the concerns over sustainable growth.

Clearly, it is up to individual countries to decide on the energy and environmental programmes best suited to their needs, within the context set by international developments and resource constraints. However, it is in the international community's interests to ensure that, so far as possible, policy choices are made on the basis of sound information and hard facts, and on a realistic appreciation of the uncertainties and risks that surround individual technologies.

It is in this respect that the international agencies, including the Nuclear Energy Agency, have played and can continue to play an important role, particularly at this critical period of global development.

Annex

NOTES FOR CHAPTER 5

1. *Uranium fission*

Natural uranium is a mixture of several isotopes. The predominant ones are uranium-238 and uranium-235, which occur in the proportions 99.3 per cent and 0.7 per cent in most natural uranium. Uranium-238 does not undergo significant fission in thermal reactors, so that only the uranium-235 is burnt to produce heat energy and fission products.

The concentration of uranium-235 in the fuel is increased by enrichment processes to around 3 per cent in the uranium oxide fuels used in light-water reactors, leaving 0.2 per cent to 0.3 per cent remaining in the depleted uranium residue. Thus only about 0.5 per cent of the initial uranium is burnable (fissionable) and available to be burnt in the reactor. Not all of this is consumed in practice however.

Capture of fission neutrons by the predominant non-fissile uranium-238 isotope converts a small proportion of it to plutonium-239 which is fissionable, and part of this plutonium is consumed *in situ* in thermal reactors, adding to the overall energy extracted from the initial uranium fuel.

2. *Discounting and rates of return on capital*

An extended discussion on discounting and discount rates is given in *Projected Costs of Generating Electricity* and its predecessors (see Bibliography). In simple terms, if money can earn r per cent per annum in real terms, $10 today will become $10(1+r)^t$ in t years time, and $10 in t years is taken to be equivalent in value to $10(1+r)^{-t}$ in present-worth terms, all in fixed value $.

The discount rate r per cent p.a. can be viewed as the opportunity cost of capital to the utility, which may be determined by market forces or by government policy. Cost comparisons can be very sensitive to the value adopted, as demonstrated in the cited NEA report.

An alternative to discounting to present values and deriving total costs is to allocate capital charges to projected electricity output. This can be done either by calculating depreciation and interest charges, which yields a variable decreasing annual sum to be recovered from the plant's output, or by annuitising the capital and interest to yield a constant sum for each year of the plant's life. The latter yields a capital charge per unit of output which, if added to the fuel and operating charges per kwh, is equal to the levelised lifetime cost derived from the total present-worth cost calculations, provided the interest on capital and discount rates employed are the same (see note below).

3. *Levelised cost calculations*

The levelised cost is derived as follows:

The expenditures on investment, operations and maintenance, and fuel (including decommissioning and spent fuel management and waste management charges) in year t are c_t, m_t and f_t respectively. The electrical output in year t is e_t and the lifetime levelised cost P. The discount rate is taken to be r per cent p.a. and the operating life of the plant L years. Then:

Total Present-Worth Cost $\quad = \quad \Sigma\ (c_t + m_t + f_t)\ (1+r)^{-t}$
summed over all time.

Total Present-Worth Output Value $\quad = \quad \Sigma\ Pe_t(1+r)^{-t}$
summed to final reactor shut-down.

For the two to be equal:

Lifetime Levelised cost P $\quad = \quad \dfrac{\Sigma\ (c_t + m_t + f_t)\ (1+r)^{-t}}{\Sigma\ e_t(1+r)^{-t}}$

4. *Reprocessing economics*

The economic criterion for break-even between MOX and uranium oxide fuel reloads in LWRs in the case where plutonium has to meet the cost of its own recovery is

$$f + g \geq (e - d)j^t$$

using the nomenclature of the plutonium fuels study *(Plutonium Fuel: An Assessment,* see Bibliography). f and g are the values attached to uranium and plutonium recovered by reprocessing spent uranium oxide fuel, while e is the cost of reprocessing (including storage, conditioning and disposal of reprocessing wastes) and d is the cost of spent fuel storage (including conditioning and disposal), both per kg heavy metal or spent fuel. j is a factor equal to $(1+r)$, where r per cent per annum is the required rate of return on capital, while t is the time delay in years between receipt of spent fuel at a reprocessing plant and the recovery of the uranium and plutonium for recycle.

LIST OF ABBREVIATIONS AND GLOSSARY OF TERMS

AGR: Advanced Gas-Cooled Reactor.

Availability Factor: The ratio of the time that the plant is available for operation in a given period of time to the total time in that period.

Baseload: The minimum load produced by an electricity network over a given period. A station used for baseload power is a station that is normally operated to provide power continuously to meet the minimum load demands.

Burn-Up: The energy obtained from nuclear fuel measured in megawatt-days per tonne.

BWR: Boiling-Water Reactor.

CANDU: Canadian Deuterium Uranium Reactor; a reactor using natural uranium fuel, cooled and moderated by pressurised heavy water.

Capacity Factor: The ratio of the total net energy generated during a period of time to the total energy generated with rated power during the same period.

CHP: Combined Heat and Power.

Decommissioning: The actions taken at the end of a facility's useful life for its planned permanent retirement from active service.

ECU: European Currency Unit.

Enrichment: Any process by which the content of a specified isotope (uranium-235, etc.) in an element is increased.

Fossil Fuel: A term applied to coal, oil and natural gas.

FR: Fast Reactor

GW: 1 Gigawatt equals 1 000 million watts.

HTR: High-Temperature Reactor.

HWR: Heavy-Water Reactor. A reactor using heavy water as moderator.

IAEA: International Atomic Energy Agency.

IEA: International Energy Agency.

KWe: Kilowatt of electrical capacity.

KWh: Kilowatt-hour. One thousand watt-hours equals 3 600 000 joules.

Levelised Cost: Levelised cost spreads total generation cost over total output to arrive at a figure which, if charged for each kWh, would exactly balance costs and income.

Load Factor: A ratio of the energy that is produced by a facility during the period considered compared to the energy that it could have produced at maximum capacity under continuous operation during the whole of that period.

LWR: Light-Water Reactor.

MWe: Unit of power capacity equal to 1 000 kWe.

Mill: A unit of currency. One-tenth of a US cent (US$0.001).

O&M: Operation and Maintenance.

Outage: Period during which a reactor is shut down.

PHWR: Pressurised Heavy-Water Reactor.

PWR: Pressurised Water Reactor.

RBMK: Soviet water-cooled graphite-moderated reactor.

Scram: Unscheduled automatic reactor shut-down.

TWh: 1 Terawatt-hour equals 1 million million watt-hours.

VVER: Soviet pressurised water reactor.

WOCA: World Outside the Centrally Planned Economies Area: comprising all countries other than the USSR, East Europe (COCOM) and China. All OECD countries are part of WOCA community. (The term, used throughout the 1980s, has been overtaken by political change.)

BIBLIOGRAPHY

Nuclear Energy Agency publications

1. *The Economics of the Nuclear Fuel Cycle,* 1985
2. *Nuclear Spent Fuel Management,* 1986
3. *Decommissioning of Nuclear Facilities; Feasibility, Needs and Costs,* 1986
4. *Nuclear Energy and its Fuel Cycle, Prospects to 2025,* 1987
5. *Advanced Water-Cooled Reactor Technologies; Rationale, State of Progress and Outlook,* 1989
6. *Projected Costs of Generating Electricity from Power Stations for Commissioning in the Period* 1995-2000 (with IEA, IAEA), 1989.
7. *Plutonium Fuel: An Assessment,* 1989
8. *Good Performance in Nuclear Projects,* 1989
9. *Uranium, Resources, Production and Demand,* 1989 (every two years)
10. *Electricity Generation from Nuclear Reactors and Uranium Demand to 2030,* 1990
11. *Means of Reducing the Capital Costs of Nuclear Power Stations,* 1990
12. *Decommissioning of Nuclear Facilities; Analysis of the Variability of Decommissioning Cost Estimates,* 1991
13. *Can Long-Term Safety be Evaluated; An International Collective Opinion* (NEA Radioactive Waste Management Committee with IAEA, EEC), 1991
14. *Nuclear Energy Data 1991* (annual publication)
15. *Small and Medium Reactors,* 1991
16. *Actinide and Fission Product Separation and Transmutation,* 1991

Nuclear Energy Agency studies in progress and forthcoming publications

1. Assessment of Demand for and Supply of Qualified Manpower
2. The Costs of Nuclear Waste Repositories
3. The Economics of the Nuclear Fuel Cycle (update)
4. The Broader Economic Impacts of Nuclear Power
5. The Spin-Off from Nuclear Power Development
6. Projected Generation Costs from Power Stations for Commissioning in the Period 2000 to 2005
7. Proceedings of October 1991 International Seminar on Decommissioning Policies
8. Plant Life Extension

MAIN SALES OUTLETS OF OECD PUBLICATIONS
PRINCIPAUX POINTS DE VENTE DES PUBLICATIONS DE L'OCDE

ARGENTINA – ARGENTINE
Carlos Hirsch S.R.L.
Galería Güemes, Florida 165, 4° Piso
1333 Buenos Aires   Tel. (1) 331.1787 y 331.2391
Telefax: (1) 331.1787

AUSTRALIA – AUSTRALIE
D.A. Information Services
648 Whitehorse Road, P.O.B 163
Mitcham, Victoria 3132   Tel. (03) 873.4411
Telefax: (03) 873.5679

AUSTRIA – AUTRICHE
Gerold & Co.
Graben 31
Wien I   Tel. (0222) 533.50.14

BELGIUM – BELGIQUE
Jean De Lannoy
Avenue du Roi 202
B-1060 Bruxelles   Tel. (02) 538.51.69/538.08.41
Telefax: (02) 538.08.41

CANADA
Renouf Publishing Company Ltd.
1294 Algoma Road
Ottawa, ON K1B 3W8   Tel. (613) 741.4333
Telefax: (613) 741.5439
Stores:
61 Sparks Street
Ottawa, ON K1P 5R1   Tel. (613) 238.8985
211 Yonge Street
Toronto, ON M5B 1M4   Tel. (416) 363.3171
Les Éditions La Liberté Inc.
3020 Chemin Sainte-Foy
Sainte-Foy, PQ G1X 3V6   Tel. (418) 658.3763
Telefax: (418) 658.3763

Federal Publications
165 University Avenue
Toronto, ON M5H 3B8   Tel. (416) 581.1552
Telefax: (416) 581.1743

CHINA – CHINE
China National Publications Import
Export Corporation (CNPIEC)
16 Gongti E. Road, Chaoyang District
P.O. Box 88 or 50
Beijing 100704 PR   Tel. (01) 506.6688
Telefax: (01) 506.3101

DENMARK – DANEMARK
Munksgaard Export and Subscription Service
35, Nørre Søgade, P.O. Box 2148
DK-1016 København K   Tel. (33) 12.85.70
Telefax: (33) 12.93.87

FINLAND – FINLANDE
Akateeminen Kirjakauppa
Keskuskatu 1, P.O. Box 128
00100 Helsinki   Tel. (358 0) 12141
Telefax: (358 0) 121.4441

FRANCE
OECD/OCDE
Mail Orders/Commandes par correspondance:
2, rue André-Pascal
75775 Paris Cedex 16   Tel. (33-1) 45.24.82.00
Telefax: (33-1) 45.24.85.00 or (33-1) 45.24.81.76
Telex: 640048 OCDE

OECD Bookshop/Librairie de l'OCDE :
33, rue Octave-Feuillet
75016 Paris   Tel. (33-1) 45.24.81.67
(33-1) 45.24.81.81

Documentation Française
29, quai Voltaire
75007 Paris   Tel. 40.15.70.00

Gibert Jeune (Droit-Économie)
6, place Saint-Michel
75006 Paris   Tel. 43.25.91.19

Librairie du Commerce International
10, avenue d'Iéna
75016 Paris   Tel. 40.73.34.60
Librairie Dunod
Université Paris-Dauphine
Place du Maréchal de Lattre de Tassigny
75016 Paris   Tel. 47.27.18.56
Librairie Lavoisier
11, rue Lavoisier
75008 Paris   Tel. 42.65.39.95
Librairie L.G.D.J. - Montchrestien
20, rue Soufflot
75005 Paris   Tel. 46.33.89.85
Librairie des Sciences Politiques
30, rue Saint-Guillaume
75007 Paris   Tel. 45.48.36.02
P.U.F.
49, boulevard Saint-Michel
75005 Paris   Tel. 43.25.83.40
Librairie de l'Université
12a, rue Nazareth
13100 Aix-en-Provence   Tel. (16) 42.26.18.08
Documentation Française
165, rue Garibaldi
69003 Lyon   Tel. (16) 78.63.32.23
Librairie Decitre
29, place Bellecour
69002 Lyon   Tel. (16) 72.40.54.54

GERMANY – ALLEMAGNE
OECD Publications and Information Centre
Schedestrasse 7
D-W 5300 Bonn 1   Tel. (0228) 21.60.45
Telefax: (0228) 26.11.04

GREECE – GRÈCE
Librairie Kauffmann
Mavrokordatou 9
106 78 Athens   Tel. 322.21.60
Telefax: 363.39.67

HONG-KONG
Swindon Book Co. Ltd.
13–15 Lock Road
Kowloon, Hong Kong   Tel. 366.80.31
Telefax: 739.49.75

ICELAND – ISLANDE
Mál Mog Menning
Laugavegi 18, Pósthólf 392
121 Reykjavik   Tel. 162.35.23

INDIA – INDE
Oxford Book and Stationery Co.
Scindia House
New Delhi 110001   Tel.(11) 331.5896/5308
Telefax: (11) 332.5993
17 Park Street
Calcutta 700016   Tel. 240832

INDONESIA – INDONÉSIE
Pdii-Lipi
P.O. Box 269/JKSMG/88
Jakarta 12790   Tel. 583467
Telex: 62 875

IRELAND – IRLANDE
TDC Publishers – Library Suppliers
12 North Frederick Street
Dublin 1   Tel. 74.48.35/74.96.77
Telefax: 74.84.16

ISRAEL
Electronic Publications only
Publications électroniques seulement
Sophist Systems Ltd.
71 Allenby Street
Tel-Aviv 65134   Tel. 3-29.00.21
Telefax: 3-29.92.39

ITALY – ITALIE
Libreria Commissionaria Sansoni
Via Duca di Calabria 1/1
50125 Firenze   Tel. (055) 64.54.15
Telefax: (055) 64.12.57
Via Bartolini 29
20155 Milano   Tel. (02) 36.50.83
Editrice e Libreria Herder
Piazza Montecitorio 120
00186 Roma   Tel. 679.46.28
Telefax: 678.47.51
Libreria Hoepli
Via Hoepli 5
20121 Milano   Tel. (02) 86.54.46
Telefax: (02) 805.28.86
Libreria Scientifica
Dott. Lucio de Biasio 'Aeiou'
Via Coronelli, 6
20146 Milano   Tel. (02) 48.95.45.52
Telefax: (02) 48.95.45.48

JAPAN – JAPON
OECD Publications and Information Centre
Landic Akasaka Building
2-3-4 Akasaka, Minato-ku
Tokyo 107   Tel. (81.3) 3586.2016
Telefax: (81.3) 3584.7929

KOREA – CORÉE
Kyobo Book Centre Co. Ltd.
P.O. Box 1658, Kwang Hwa Moon
Seoul   Tel. 730.78.91
Telefax: 735.00.30

MALAYSIA – MALAISIE
Co-operative Bookshop Ltd.
University of Malaya
P.O. Box 1127, Jalan Pantai Baru
59700 Kuala Lumpur
Malaysia   Tel. 756.5000/756.5425
Telefax: 757.3661

NETHERLANDS – PAYS-BAS
SDU Uitgeverij
Christoffel Plantijnstraat 2
Postbus 20014
2500 EA's-Gravenhage   Tel. (070 3) 78.99.11
Voor bestellingen: Tel. (070 3) 78.98.80
Telefax: (070 3) 47.63.51

NEW ZEALAND
NOUVELLE-ZÉLANDE
Legislation Services
P.O. Box 12418
Thorndon, Wellington   Tel. (04) 496.5652
Telefax: (04) 496.5698

NORWAY – NORVÈGE
Narvesen Info Center – NIC
Bertrand Narvesens vei 2
P.O. Box 6125 Etterstad
0602 Oslo 6   Tel. (02) 57.33.00
Telefax: (02) 68.19.01

PAKISTAN
Mirza Book Agency
65 Shahrah Quaid-E-Azam
Lahore 3   Tel. 66.839
Telex: 44886 UBL PK. Attn: MIRZA BK

PORTUGAL
Livraria Portugal
Rua do Carmo 70-74
Apart. 2681
1117 Lisboa Codex   Tel.: (01) 347.49.82/3/4/5
Telefax: (01) 347.02.64

SINGAPORE – SINGAPOUR
Information Publications Pte. Ltd.
41, Kallang Pudding, No. 04-03
Singapore 1334   Tel. 741.5166
                               Telefax: 742.9356

SPAIN – ESPAGNE
Mundi-Prensa Libros S.A.
Castelló 37, Apartado 1223
Madrid 28001   Tel. (91) 431.33.99
                          Telefax: (91) 575.39.98
Libreria Internacional AEDOS
Consejo de Ciento 391
08009 – Barcelona   Tel. (93) 488.34.92
                          Telefax: (93) 487.76.59
Llibreria de la Generalitat
Palau Moja
Rambla dels Estudis, 118
08002 – Barcelona
          (Subscripcions) Tel. (93) 318.80.12
          (Publicacions) Tel. (93) 302.67.23
                          Telefax: (93) 412.18.54

SRI LANKA
Centre for Policy Research
c/o Colombo Agencies Ltd.
No. 300-304, Galle Road
Colombo 3   Tel. (1) 574240, 573551-2
                Telefax: (1) 575394, 510711

SWEDEN – SUÈDE
Fritzes Fackboksföretaget
Box 16356
Regeringsgatan 12
103 27 Stockholm   Tel. (08) 690.90.90
                          Telefax: (08) 20.50.21
Subscription Agency-Agence d'abonnements
Wennergren-Williams AB
Nordenflychtsvägen 74
Box 30004
104 25 Stockholm   Tel. (08) 13.67.00
                          Telefax: (08) 618.62.36

SWITZERLAND – SUISSE
Maditec S.A. (Books and Periodicals - Livres
et périodiques)
Chemin des Palettes 4
1020 Renens/Lausanne   Tel. (021) 635.08.65
                          Telefax: (021) 635.07.80

Librairie Payot S.A.
4, place Pépinet
1003 Lausanne   Tel. (021) 341.33.48
                          Telefax: (021) 341.33.45

Librairie Unilivres
6, rue de Candolle
1205 Genève   Tel. (022) 320.26.23
                          Telefax: (022) 329.73.18

Subscription Agency - Agence d'abonnement
Naville S.A.
38 avenue Vibert
1227 Carouge   Tél.: (022) 308.05.56/57
                          Telefax: (022) 308.05.88

See also – Voir aussi :
OECD Publications and Information Centre
Schedestrasse 7
D-W 5300 Bonn 1 (Germany)
                           Tel. (49.228) 21.60.45
                          Telefax: (49.228) 26.11.04

TAIWAN – FORMOSE
Good Faith Worldwide Int'l. Co. Ltd.
9th Floor, No. 118, Sec. 2
Chung Hsiao E. Road
Taipei   Tel. (02) 391.7396/391.7397
                          Telefax: (02) 394.9176

THAILAND – THAÏLANDE
Suksit Siam Co. Ltd.
113, 115 Fuang Nakhon Rd.
Opp. Wat Rajbopith
Bangkok 10200   Tel. (662) 251.1630
                          Telefax: (662) 236.7783

TURKEY – TURQUIE
Kültur Yayinlari Is-Türk Ltd. Sti.
Atatürk Bulvari No. 191/Kat. 13
Kavaklidere/Ankara   Tel. 428.11.40 Ext. 2458
Dolmabahce Cad. No. 29
Besiktas/Istanbul   Tel. 160.71.88
                            Telex: 43482B

UNITED KINGDOM – ROYAUME-UNI
HMSO
Gen. enquiries   Tel. (071) 873 0011
Postal orders only:
P.O. Box 276, London SW8 5DT
Personal Callers HMSO Bookshop
49 High Holborn, London WC1V 6HB
                          Telefax: (071) 873 8200
Branches at: Belfast, Birmingham, Bristol, Edin-
burgh, Manchester

UNITED STATES – ÉTATS-UNIS
OECD Publications and Information Centre
2001 L Street N.W., Suite 700
Washington, D.C. 20036-4910 Tel. (202) 785.6323
                          Telefax: (202) 785.0350

VENEZUELA
Libreria del Este
Avda F. Miranda 52, Aptdo. 60337
Edificio Galipán
Caracas 106   Tel. 951.1705/951.2307/951.1297
                      Telegram: Libreste Caracas

Subscription to OECD periodicals may also be
placed through main subscription agencies.

Les abonnements aux publications périodiques de
l'OCDE peuvent être souscrits auprès des
principales agences d'abonnement.

Orders and inquiries from countries where Distributors have not yet been appointed should be sent to:
OECD Publications Service, 2 rue André-Pascal,
75775 Paris Cedex 16, France.

Les commandes provenant de pays où l'OCDE n'a
pas encore désigné de distributeur devraient être
adressées à : OCDE, Service des Publications,
2, rue André-Pascal, 75775 Paris Cedex 16, France.

                                        12-1992

QMW LIBRARY
(MILE END)

OECD PUBLICATIONS, 2 rue André-Pascal, 75775 PARIS CEDEX 16
PRINTED IN FRANCE
(66 92 15 1) ISBN 92-64-13798-X - No. 46121 1992